ENGINEERING IN PLAIN SIGHT

ENGINEERING IN PLAIN SIGHT

AN ILLUSTRATED FIELD GUIDE TO THE CONSTRUCTED ENVIRONMENT

by Grady Hillhouse

no starch press

San Francisco

Printed in China

Second printing

27 26 25 24 23   2 3 4 5 6

ISBN-13: 978-1-7185-0232-1 (print)
ISBN-13: 978-1-7185-0233-8 (ebook)

Publisher: William Pollock
Managing Editor: Jill Franklin
Production Manager: Rachel Monaghan
Developmental Editor: Jill Franklin
Production Editor: Rachel Monaghan
Illustration and Cover Design: MUTI
Technical Reviewers: Thomas Overbye, Robert Weller, Laurence Rillet, Brian Gettinger, John Sobanjo, Erol Tutumluer, Tina McMartin, Jennifer Elms, and Brandon White
Interior Design and Composition: Maureen Forys, Happenstance Type-O-Rama
Proofreader: Audrey Doyle

For information on distribution, bulk sales, corporate sales, or translations, please contact No Starch Press, Inc. directly at info@nostarch.com or:

No Starch Press, Inc.
245 8th Street, San Francisco, CA 94103
phone: 1.415.863.9900
www.nostarch.com

Library of Congress Control Number: 2022013814

[I]

To Crystal

Contents

INTRODUCTION

In mid-2009, as the world was just escaping the most severe economic meltdown since the 1930s, I was escaping university with an undergraduate liberal arts degree and not a single prospect for gainful employment. Rather than take my chances in the dreadful job market, I decided to spend a little more time (and a lot more money) on my education. Facing the difficult reality that a university degree could not guarantee a job, I diligently cross-referenced my various interests against their occupational prospects to refocus my career path in a more dependable and less ambiguous direction. I settled on civil engineering, a subject about which I knew almost nothing but that seemed both exciting and responsible. Incredibly, I was accepted into my top choice for graduate school and started my studies that fall.

Once I had worked through the basic math and science classes required to catch up to my graduate-level classmates, I began the engineering coursework. I have always been generally curious about science, technology, and how things work. Still, nothing could have prepared me for the remarkable transformation of perspective I would receive during the rest of my studies. Structural design classes had me staring at every beam and column to be spotted in each new building I visited. Circuits labs pointed out the details and complexities of electric transmission lines and substations. Stormwater engineering lectures compelled me to notice every drain, manhole, channel, and detention basin while biking or driving around town. Each and every class was like turning on a lamp to illuminate some innocuous part of the constructed environment that I had never noticed before. I was captivated.

I finished my degree not only with a job but also with an entirely new way of looking at the world. It didn't take long for that enthusiasm and excitement about infrastructure to overflow into my personal life, including my hobby YouTube channel. What started as a way to share my woodworking projects with other makers and craftspeople slowly turned into an outlet for introducing topics in engineering to the world. Now I produce educational videos full-time, and *Practical Engineering* has millions of viewers each month.

Even the most unexceptional parts of the built environment are monuments to the solutions to hundreds of practical engineering problems. Understanding even a small subset of those challenges and their resolutions had the power to instill astonishment and wonder in me, and I never stopped feeling that way. Now my entire life is essentially a treasure hunt for all the interesting little details of the constructed world. My spontaneous stops at each dam and bridge for a photograph or a better look drive my wife a little crazy on road trips. I regularly lose my train of thought on walks when noticing some new or different piece of infrastructure. And there is a tiny part of my brain that is dedicated—no matter where I am or what I'm doing—to following the path that

stormwater will take as it runs off along the ground. Engineering opened my eyes to the infrastructure that surrounds and supports our modern lives. If any of that eagerness comes across in this book, I have succeeded.

This is not a comprehensive field guide. Infrastructure takes a myriad of shapes and forms that vary around the world. This book focuses on the United States, but constructed works can look quite different even between states, counties, and cities. It wouldn't be practical to try to document them all. Plus, it would ruin the fun. Part of the joy of "infrastructure spotting" is using your detective skills to deduce the purpose of random bits and bobs as you come across them. I hope what follows can ignite that joy and further your journey as an enthusiastic beholder of the constructed environment.

—Grady Hillhouse

1

ELECTRICAL GRID

Introduction

Harnessing the power of electricity is one of humanity's greatest achievements. What was a luxury 100 years ago is now a critical resource for the safety, prosperity, and well-being of nearly everyone. In the not-too-distant past, manpower and horsepower were practically the only powers. Hard work was accomplished through the strength of living beings. It's no wonder we humans have sought to take control of energies beyond our own bodies. These days, "energy" gives life to almost every aspect of the contemporary world, enabling our most basic physiological needs to our most cutting-edge technologies.

Depending on how it is harnessed, stored, distributed, and used, energy can take many forms. On the Earth, we can trace nearly all our energy back to the sun. Wind and waves are created by heating of the Earth's atmosphere. Solar light can be harnessed directly. Even fossil fuels like gasoline got their energy from the sun. Prehistoric plants captured this solar power through photosynthesis and were buried over millions of years, only to be tapped into by wells, extracted, refined, and exploded in engines, releasing the sun's heat (along with many other foul byproducts) back to the planet again. Humans do a lot of converting of energy from one form to another for convenience and practicality, but nothing compares to electricity, which makes having a personal source of power possible for nearly everyone.

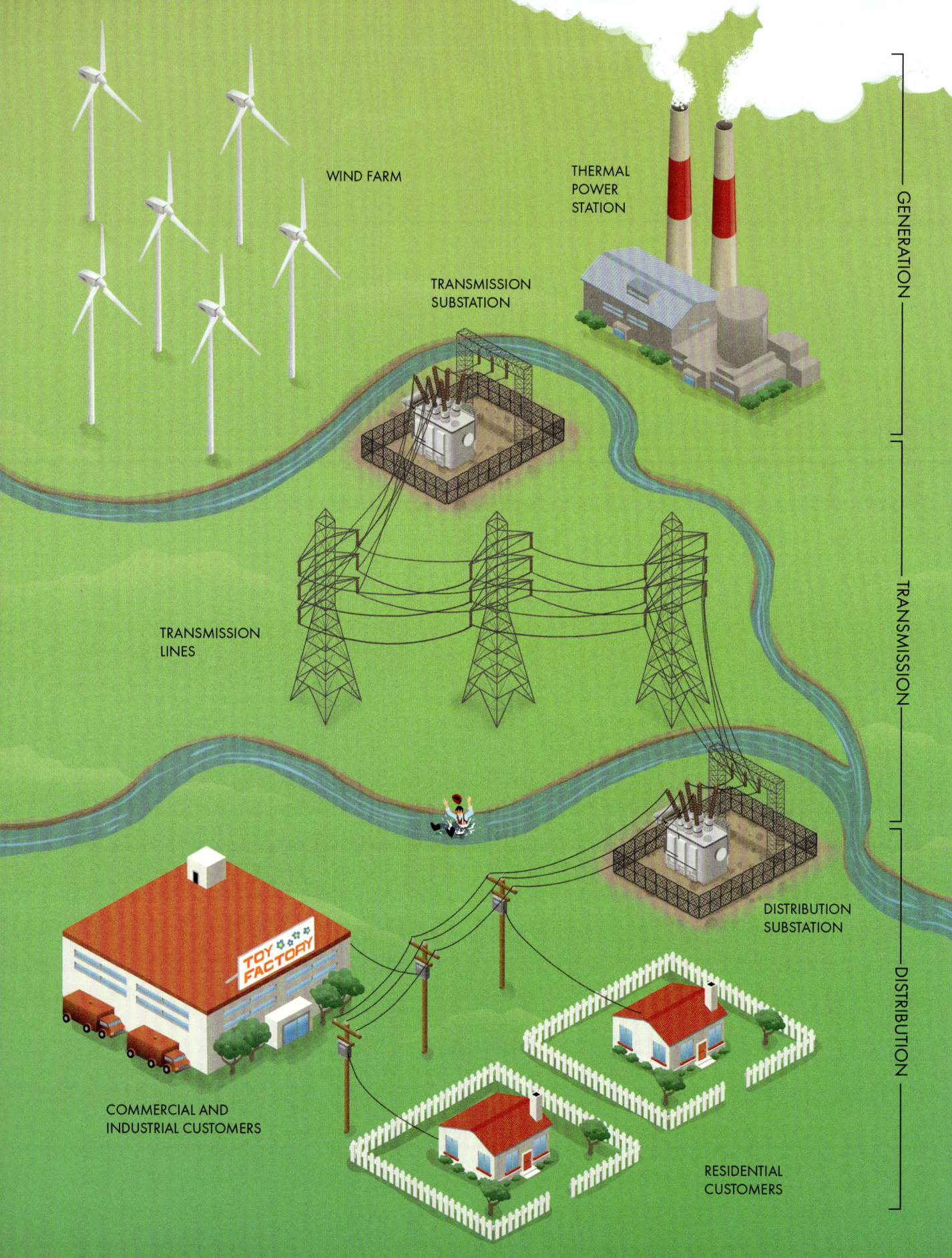

WIND FARM

THERMAL
POWER
STATION

TRANSMISSION
SUBSTATION

TRANSMISSION
LINES

DISTRIBUTION
SUBSTATION

TOY
FACTORY

COMMERCIAL AND
INDUSTRIAL CUSTOMERS

RESIDENTIAL
CUSTOMERS

GENERATION

TRANSMISSION

DISTRIBUTION

Overview of the Electrical Grid

Electricity is remarkably different from all the other types of energy. We can't hold it in our hands, and we can't see it directly. Yet, it can accomplish incredible work—from physical feats to computations—nearly instantaneously. Rather than being a tangible manifestation of energy, such as fuels, electricity takes a more transient form, requiring only a connection by metal wires for transmission. The simplicity of moving it from one place to the next has given rise to *electrical grids*, huge, interconnected networks of electricity producers and users. To get a sense of scale, only five major power grids cover all of North America, and many of the world's largest power grids encompass multiple countries.

In general, electricity makes its way through a series of discrete steps on the grid divided into three parts: GENERATION (production of power), TRANSMISSION (moving that electricity from centralized plants to populated areas), and DISTRIBUTION (delivering the electricity to every individual customer). SUBSTATIONS serve as the connection points between the major parts. Establishing these large interconnections solves a lot of challenges at once. Allowing a greater number of users and producers to share expensive infrastructure creates efficiency. Because power can take many different paths to each location, and individual power plants can step in if another one falls offline, reliability increases. Finally, interconnections help smooth out the flow of electricity.

Unlike other utilities, electricity is quite challenging to store on a large scale, which means power must be generated, transported, supplied, and used all in the same moment. The energy coursing through the wires of your home or office was a ray of sunshine on a solar panel, an atom of uranium, or a bit of coal or natural gas in a steam boiler only milliseconds ago. The electricity a single household uses can be quite sporadic. The more users that can be connected together, the more everyone's spikes and swings in usage average out.

Making a gigantic, one-size-fits-all electrical grid work for every type of power user and producer is no simple feat. You can imagine the power grid as a freight train going up a hill, with locomotives representing generation and freight representing electrical demands. All the engines must move in perfect synchrony to share the load. If one is slower or faster than the rest, it runs the risk of breaking the whole train. To make it even more challenging, the demands on the grid are continually changing over time like valleys and hills in the landscape. Power consumers turn electrical devices on and off at will, with no notification to the utilities. Demands peak during the day when people are using lots of electricity, particularly on scorching or freezing days when many are using air conditioners or heaters. To avoid *brownouts* and *blackouts*, generation must be continuously adjusted up or down to match electrical demands on the grid. This process is called *load following*, just like a locomotive adjusts its throttle to account for changes in grade along the way.

Electricity customers use power in different ways. COMMERCIAL AND INDUSTRIAL CUSTOMERS adjust their usage based on the fluctuating price of electricity, often running machines overnight to take advantage of the cheaper energy. RESIDENTIAL CUSTOMERS (who normally pay a fixed price) can be less attentive to the ebbs and swings of total grid demands, using electricity whenever it is most convenient.

Similarly, different types of power plants are able to generate electricity in different ways. Solar farms generate lots of electricity when the sun is up, but none when the sun is down. WIND FARMS generate electricity depending on weather, with peak output during times when winds are strong and consistent. Nuclear plants generate consistent power with little ability to ramp up or down, while other THERMAL POWER STATIONS like coal or natural gas plants can adjust their output somewhat according to changing demands. Hydropower plants are the most responsive, often with the ability to start and stop generation within seconds or minutes.

Grid managers perform detailed forecasts of both generation and demand to make sure they can maintain balance between the two. They have to consider when to schedule outages of power plants and TRANSMISSION LINES for maintenance and quickly adjust when facilities trip offline without notice due to damage or other issues. They hope for the best but plan for the worst, taking into account the abilities and limitations of the entire portfolio of power producers and users. If the worst comes, and there is not enough electricity to meet demands, grid managers will require that some customers be temporarily disconnected (called *load shedding*) to reduce demands and avoid a total collapse. Normally these disconnections happen on a rolling basis of 15 to 30 minutes to spread out the inconvenience of lost service, so they are often known as *rolling outages*.

Many types of equipment are needed to generate, transmit, and deliver electricity over large areas. Remarkably, most of this infrastructure is out in the open for anyone to have a look. Many times, I've been accused of having my head in the clouds when I was just observing something at the top of a utility pole. You can examine and identify nearly every major piece of the electrical grid no matter where you are. The rest of this chapter provides a closer look at each part of the grid and more detail about the equipment and processes needed to keep the flow of electricity moving.

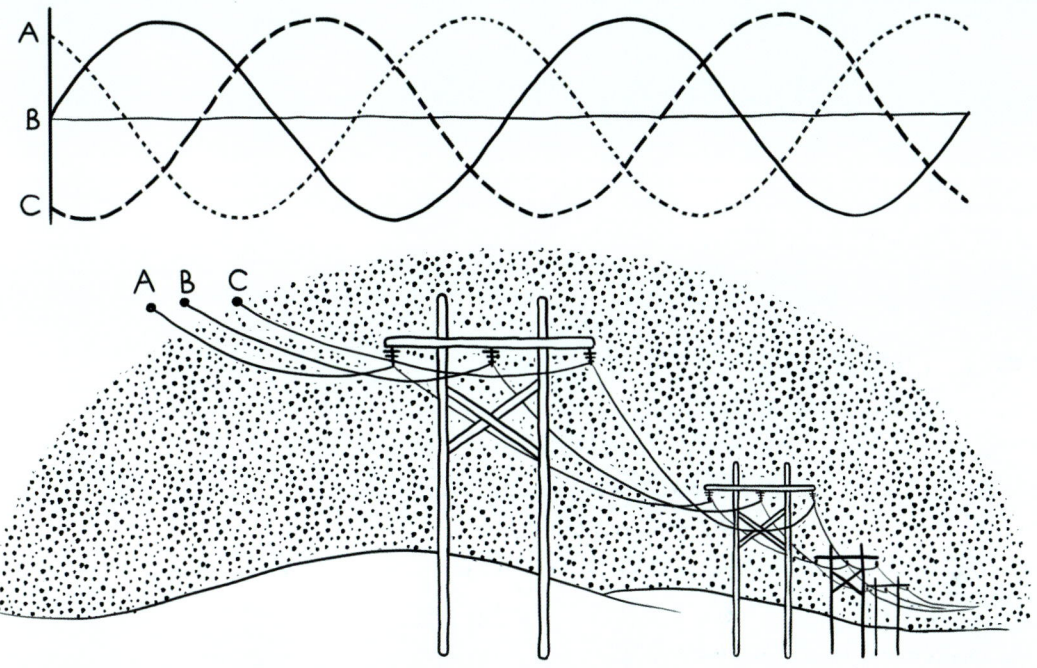

Rather than a constant flow of current in a single direction (called *direct current* or *DC*), the vast majority of the power grid uses *alternating current* or *AC*, where the direction of voltage and current is continually switching. The benefit of having the current alternate is that its voltage can easily be stepped up or down using a transformer. In North America, this happens at 60 cycles per second, giving electrical infrastructure that familiar low hum. Power is usually generated and transmitted on three individual lines called *phases* (sometimes labeled A, B, and C), each of which has voltage offset from its neighbors. Creating electricity in three distinct phases provides a smooth supply that overlaps, so there is never a moment when all phases have zero voltage. A three-phase supply also uses fewer equivalent conductors than a single-phase supply to carry the same amount of power, making it more economical. You'll notice that almost all electrical infrastructure shows up in groups of three, with each conductor or piece of equipment handling an individual phase of the supply.

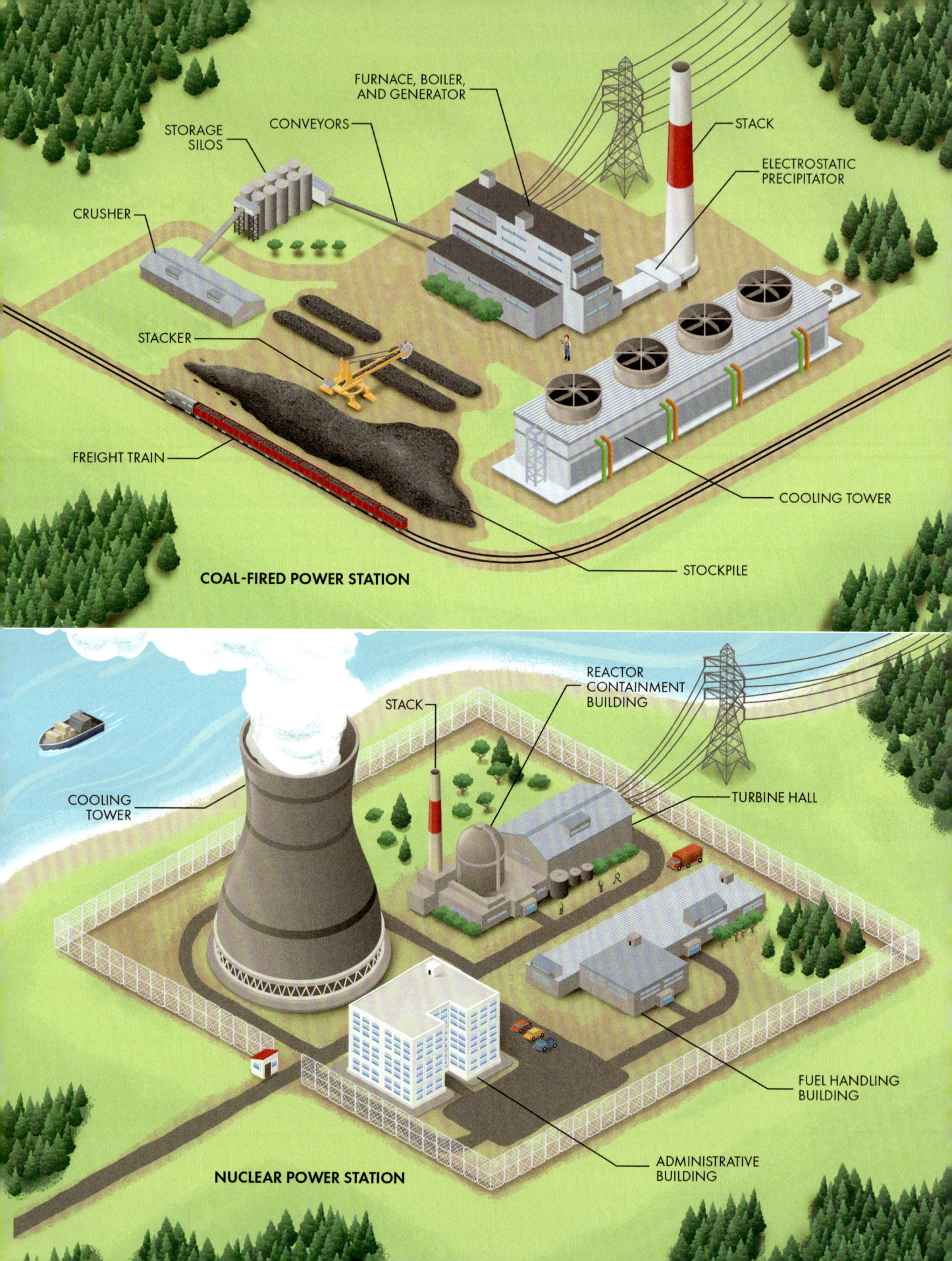

CRUSHER

STORAGE SILOS

CONVEYORS

FURNACE, BOILER, AND GENERATOR

STACK

ELECTROSTATIC PRECIPITATOR

STACKER

FREIGHT TRAIN

COOLING TOWER

STOCKPILE

COAL-FIRED POWER STATION

COOLING TOWER

STACK

REACTOR CONTAINMENT BUILDING

TURBINE HALL

FUEL HANDLING BUILDING

ADMINISTRATIVE BUILDING

NUCLEAR POWER STATION

Thermal Power Stations

Generation is the first step electricity takes on its journey through the power grid, a trip that may be hundreds or thousands of miles, but that happens almost instantly. Although most of us do not have a power plant in our backyard, we do have an immediate link to each one connected to the grid. There are many types of power plants, each with distinct advantages and disadvantages, but they all have one thing in common: they take some kind of energy that can be harvested from the natural environment and convert it into electrical energy for use on the grid. Many of the methods we use to generate power are just different ways of boiling water. Plants that use this method are called *thermal power stations* because they rely on heat to create steam. The steam passes through a turbine, which is coupled to an AC GENERATOR connected to the power grid. The speed of the turbine must be carefully synchronized to the frequency of the rest of the grid.

Most power plants are sophisticated industrial facilities closed to visitors. In fact, be careful not to lurk suspiciously nearby because many plants are heavily guarded! However, you can still spot them regularly from highways or airplane windows by keeping an eye out for large congregations of high-voltage transmission lines and the recognizable tall stacks. Pay special attention to lakes outside large cities as well, because they sometimes serve as a source of cooling water for power plants. A detailed explanation of how thermal power plants work is beyond the scope of this book, but there is a lot of satisfaction in seeing and understanding the parts and pieces that you can observe from the outside.

A large amount of our electricity starts as fossil fuels (mainly coal or natural gas). COAL-FIRED POWER STATIONS are becoming less common as other fuels grow less expensive and, more important, are less polluting. However, coal still makes up a large proportion of overall electricity generation. You'll know immediately if you've spotted a coal power station, because most of the visible infrastructure will be related to handling the coal itself. These plants process and burn thousands of tons of fuel every day, so they need lots of equipment to offload, store, crush, and transport the coal to the FURNACE and BOILER.

Unless the plant is situated next door to a coal mine, the primary way to move this much fuel efficiently is by FREIGHT TRAIN. Complex systems of railways often surround these plants to allow for frequent and efficient delivery of fuel. Trucks and barges are sometimes used to deliver fuel when railways aren't feasible. Coal STACKERS are massive moveable conveyor belts for bulk handling of the fuel. They travel on tracks and use their booms to organize and create STOCKPILES of coal. Plants normally maintain several weeks' worth of fuel to make sure they can continue to operate if the supply is temporarily disrupted.

Unlike a charcoal grill in your backyard, most power plant furnaces burn a constant stream of fine coal powder. Coal is delivered in large chunks, so from the stockpiles, it must go into a CRUSHER to reduce its size for more efficient burning. Between each step of the fuel handling, large covered CONVEYOR BELTS transport the coal. STORAGE SILOS protect the crushed coal from the elements. From there, it makes its final journey to the furnace and boiler.

Natural gas–fired power stations (not shown here) can be identified by a lack of this coal handling equipment. Gas pipelines that feed these plants are usually underground, hidden from view, which means gas plants usually appear much simpler and smaller from the outside. For both coal- and gas-fired plants, the air from the combustion of fossil fuels is called *flue gas*, and it can carry dangerous pollutants like ash and nitrous oxides. Environmental regulations require that flue gas be rid of the worst of these pollutants before it's released into the atmosphere since they can be harmful to humans and animals. Many different facilities are used to remove pollutants from flue gas, including *baghouses* that use fabric filters, ELECTROSTATIC PRECIPITATORS that capture particles through static cling, and *scrubbers* that clean the air by spraying a fine mist, catching dust and ash. After passing through these facilities, the flue gas can be released through a STACK. Although these tall chimneys don't clean the flue gas directly, they do help manage pollution by releasing it high enough to be dispersed in the air (since dilution is sometimes the solution to pollution).

One type of thermal plant does not rely on combustion of fuel. Instead, NUCLEAR POWER STATIONS rely on the carefully controlled *fission* of radioactive materials. This process happens in a nuclear REACTOR, often evident from the outside as a pressurized CONTAINMENT BUILDING with a domed roof. The reactor building usually has an outside armoring layer of thick concrete as a precaution against natural disasters or sabotage. A separate FUEL HANDLING BUILDING is generally used for receipt, inspection, and storage of nuclear fuel. Offices and controls are often located in an ADMINISTRATIVE BUILDING, away from the fuel and equipment. Nuclear plants sometimes also have a STACK, but it is not for releasing flue gas. In some reactors, the water used to drive the TURBINE comes into direct contact with radioactive fuel, which can create gases like hydrogen and oxygen that become mildly radioactive themselves. The tall, solitary stack seen at some nuclear plants allows for safe ventilation of those gases.

The iconic symbol of a nuclear plant is the COOLING TOWER emitting unidentified plumes of gas. In reality, this gas is just water vapor. Nearly all thermal power stations use cooling towers. A separate stream of water is needed to condense the steam after it has passed through the turbine. After this water has absorbed so much heat, it can't be immediately released back into the environment because hot water is harmful to aquatic wildlife. Special structures are used to cool the water before it can be discharged or reused. The familiar wide concrete chimneys are open around the bottom to use natural drafts for cooling. Shorter, boxier units use fans. In both cases, you may be able to see the spray of water raining down along the bottom of the tower to help with evaporation.

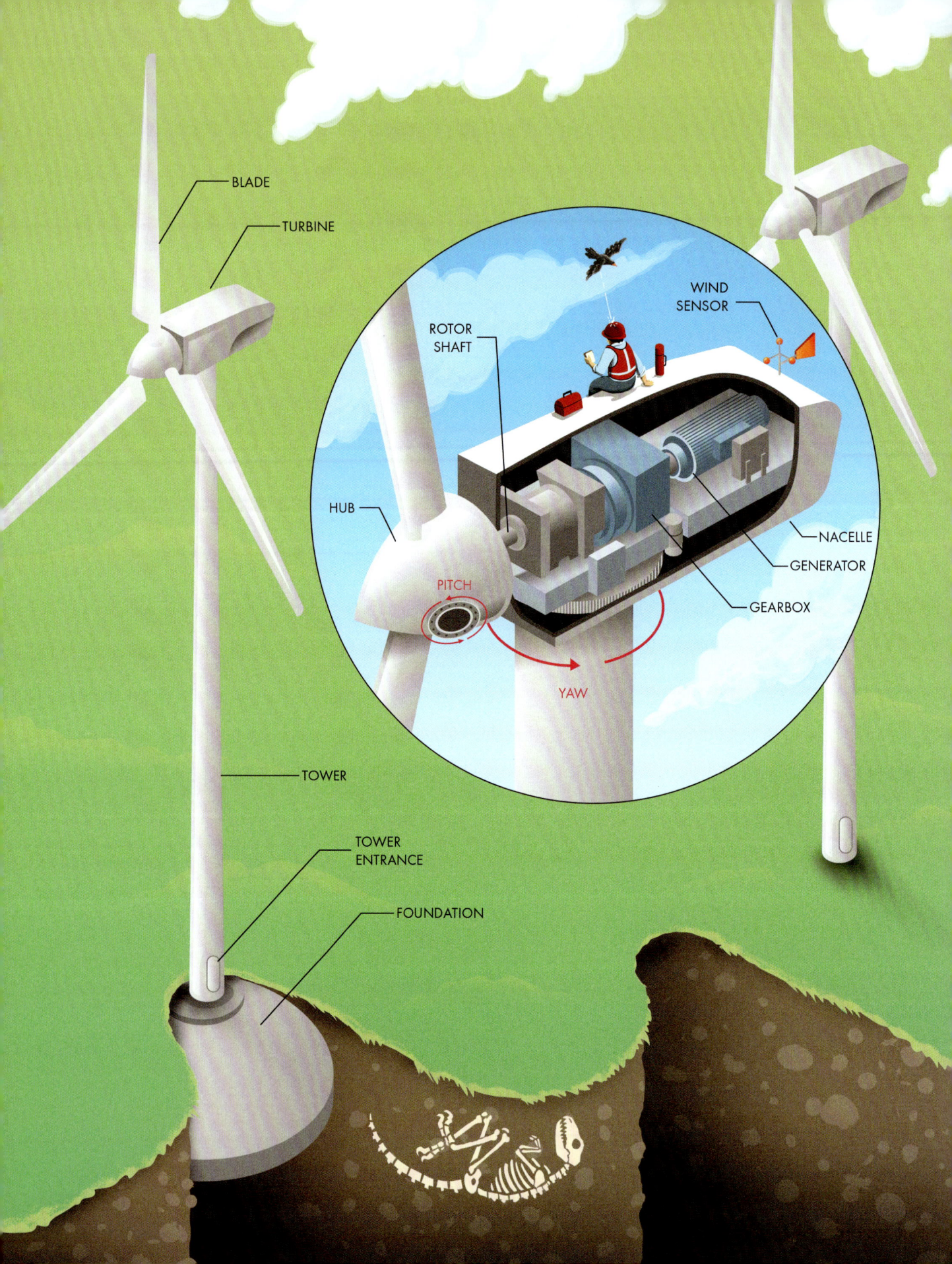

Wind Farms

Wind farms consist of multiple turbines that capture wind energy and convert it into electricity. In a way, they harvest solar energy, because wind currents are driven by the heating and cooling of the atmosphere by the sun. Since we can't choose when the wind will blow, wind farms are less reliable than thermal power plants. Grid operators in areas with lots of wind turbines must rely on weather forecasts not only to predict electrical usage, but also to predict electricity production. However, unlike coal, natural gas, and uranium, wind is free and it's going to blow whether we have turbines to harvest its power or not. Taking advantage of such a resource only makes sense, and modern wind farms have become a relatively low-cost and low-pollution part of our energy portfolio.

Wind turbines come in a wide variety of shapes and sizes, but modern variants around the world have converged to a consistent, instantly recognizable style. This design features a horizontal axis TURBINE atop a tall steel TOWER with three slender composite BLADES, all usually colored in pure white for visibility. If you didn't know better, you might assume they were modern art pieces dotting the landscape, somehow appearing both sleek and ungainly at once. Towers are usually attached to a massive concrete FOUNDATION buried below the ground and are almost always hollow with an ENTRANCE at the bottom for maintenance workers and a ladder up to the turbine. The foundation is designed to prevent the tower from toppling, even under the most extreme wind conditions.

Utility-scale turbines are usually rated around 1 to 2 megawatts each, but units as large as 10 megawatts have been installed. That's enough to power about 5,000 households with a single turbine! From the outside, you can see the HUB with attached blades and the NACELLE, the outer housing for the rest of the turbine's equipment. Inside the nacelle are the ROTOR SHAFT, GEARBOX, GENERATOR, and other equipment.

Every aspect of a turbine is intended to capture as much of the energy from the wind as possible. An important part of a turbine's efficiency is how fast the blades rotate. If they go too slowly, wind will pass through the gaps in the blades without providing any power. If they spin too quickly, the blades will block the wind, reducing the amount of power that can be harvested. I remember taking a tour of a wind farm as a kid and trying to race the shadow of the blades on the ground. I would move toward the hub's shadow little by little until I could keep up with the rate of rotation. It turns out that a turbine is most efficient when the tip of the blades is moving around four to seven times the speed of the wind. Since larger turbines have longer blades, they rotate slower to keep the tip speed near this ideal range. Even though these blades seemed quite fast to me as a child, electrical generators need to spin much faster to operate efficiently and keep up with the alternating frequency of the grid. Most turbines use a gearbox to convert the slow pace of the blades to a speed more suitable for the generator.

Turbines operate at their best when facing directly into the wind. Older windmills used a large tail to keep this proper orientation, called YAW. Modern turbines use a WIND SENSOR atop the nacelle to measure both the speed and direction of the wind. If the wind vane senses a change in direction, it directs motors to adjust the yaw of the turbine back into the wind. Most turbines also include a way to adjust the angle, or PITCH, of each blade. When the wind is too fast for the turbine to operate efficiently, the blades are furled (that is, tilted so only the edge of the blade faces into the wind) to reduce the forces on the turbine. You may wonder why, during a very windy day or a storm, all the turbines in a wind farm have stopped turning. In extreme winds or emergencies, operators apply a mechanical brake to stop the rotation and prevent damage to the equipment.

Another aspect of a wind turbine's efficiency is the narrow shape of the blades. You might think that a wider blade would allow more wind energy to be harvested, but consider this: if 100 percent of the power could be extracted from the wind, the air would have no velocity left to exit behind the blades. This would cause air to "pile up" and block any new wind from driving the turbine. Some wind movement is required to keep fresh air supplying the turbine, which means it's never possible to harvest all the energy from the wind. The theoretical maximum efficiency that can be extracted (called the *Betz limit*) is about 60 percent. The slender blades of the turbine are carefully designed to capture as much energy as possible without slowing the air stream too much.

If you drive past or fly over a wind farm at night, you'll notice the red lights on top of the towers. Like on all tall towers and buildings, these lights are a warning to aircraft to help avoid collisions. On most wind farms, the warning lights flash in perfect synchrony to help aircraft pilots judge the shape and extent of the entire wind farm. If all the lights blinked randomly, it would be too disorienting. Maintaining this synchronicity among all the turbines in a wind farm is a totally separate challenge. You might think that all the lights would need to be wired together, but the complexity of such a system would be unreliable and costly. Instead, each light is outfitted with a GPS receiver that gets a highly accurate clock signal from satellites overhead. If each light has its clock synchronized, the flashes will be synchronized as well.

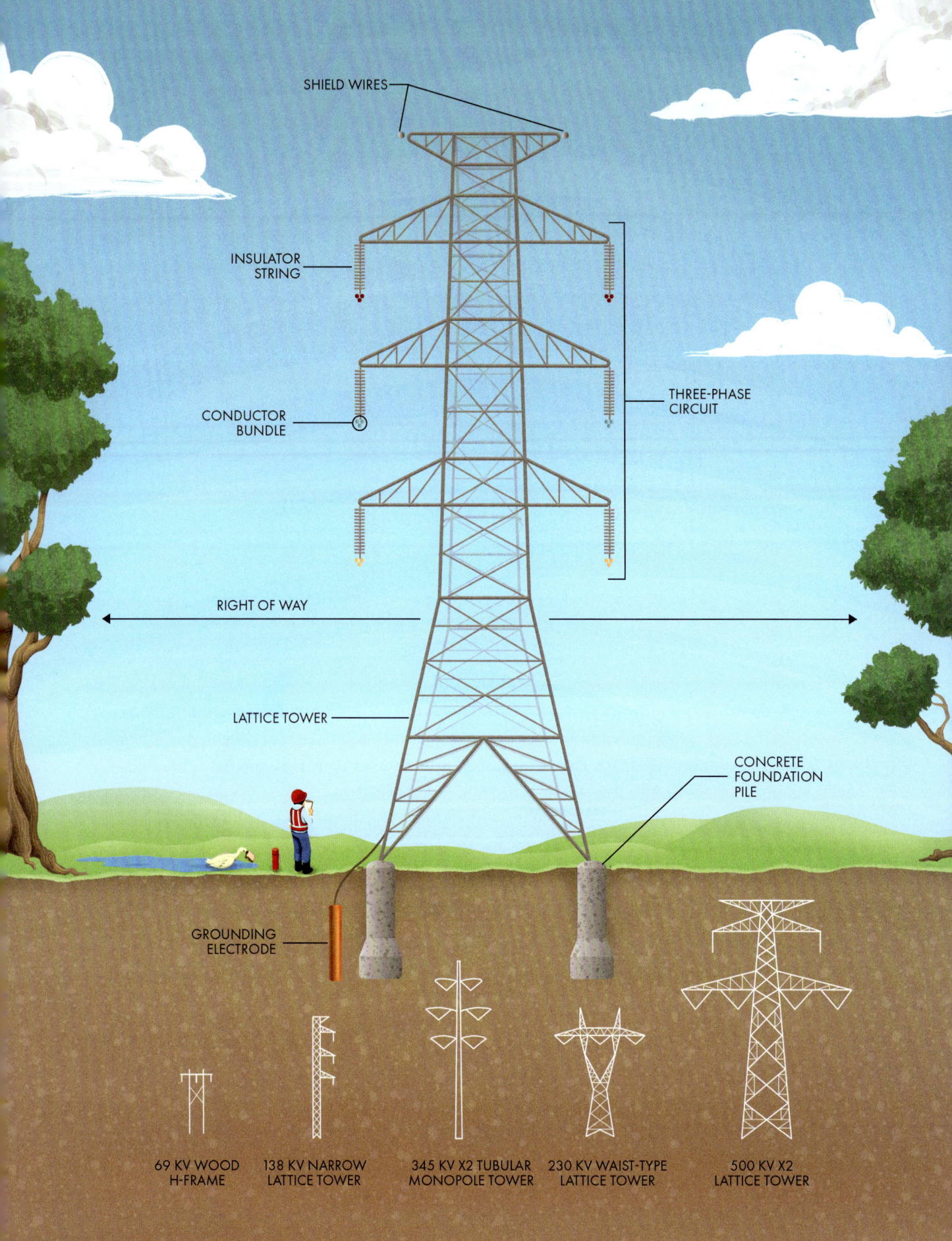

SHIELD WIRES

INSULATOR STRING

CONDUCTOR BUNDLE

THREE-PHASE CIRCUIT

RIGHT OF WAY

LATTICE TOWER

CONCRETE FOUNDATION PILE

GROUNDING ELECTRODE

69 KV WOOD H-FRAME

138 KV NARROW LATTICE TOWER

345 KV X2 TUBULAR MONOPOLE TOWER

230 KV WAIST-TYPE LATTICE TOWER

500 KV X2 LATTICE TOWER

Transmission Towers

Power plants are almost always located far away from populated regions. The land is cheaper in rural areas, and most people don't like to live near huge industrial facilities. It just makes sense to keep some distance between our power plants and cities. However, creating all the electricity far from where it's needed presents a transportation challenge. You can't load electricity onto trucks and deliver it to customers. Instead, it travels instantaneously from producers to users on wires we call transmission lines. You may be familiar enough with this concept if you've ever used extension cords to bring power to lights or devices that can't reach an electrical outlet. Scaling up this operation for the bulk transmission of electricity from power plants, however, creates some interesting challenges.

Wires used for the transmission of electricity are called CONDUCTORS, and no conductor is perfect. You can put electricity in on one side, but you never get 100 percent of it out at the end. That's because all conductors have some *resistance* to the flow of electricity. This resistance converts some of the electricity to heat, wasting its power along the way. Generating electricity is a costly and complex process, so if we're going to go to all that trouble, we want to make sure that as much of it as possible actually reaches the customers for whom it's intended. Luckily, there's a trick to reducing the amount of energy that gets wasted from the resistance of transmission lines, but it requires understanding a little bit about electrical circuits.

Electricity flowing in a circuit has two important properties: *voltage* is the amount of electric potential (somewhat equivalent to the pressure of a fluid in a pipe), and *current* is the flow rate of an electric charge (like the flow rate of a fluid in a pipe). These two properties are related to the total amount of power traveling through a line. The amount of wasted power from resistance is related to the current in the line, so more current means more wastage. If you increase the voltage of the electricity, you need less current to deliver the same amount of power, so that's exactly what we do. Transformers at power plants boost the voltage before sending electricity on its way over transmission lines, which reduces the current in the lines, minimizes energy wasted due to resistance of the conductors, and ensures that as much power as possible reaches the customers at the other end.

These high voltages make electrical transmission much more efficient, but they create a new set of challenges. High voltage is extremely dangerous, so conductors need to be kept far away from human activity on the ground. Running high-voltage transmission lines underground is quite expensive, so they're typically strung overhead on TOWERS (also called *pylons*) except in the most densely populated areas.

There are many factors to consider in the design of an electrical transmission line, leading to a massive variety of shapes, sizes, and materials used for these towers. One of the most fundamental of those is the voltage of the line. The higher the voltage, the more distance required between each PHASE and above the ground. Many transmission lines carry multiple THREE-PHASE CIRCUITS to save cost, so instead of three phases, you may see six or even nine. The illustration shows only a few examples of the unique shapes and sizes of towers that can be built.

The width of the RIGHT OF WAY is also important. In urban areas, land is more expensive, so the available width to run transmission lines may be much smaller than for lines run across rural areas. A narrower path means arranging conductors vertically rather than horizontally, increasing the height (and cost) of the towers. Finally, there are aesthetic considerations. I find transmission towers to be interesting and beautiful. However, many people believe these towers create an imposition on the landscape, and they are sometimes considered a type of visual pollution. People generally tend to prefer the look of MONOPOLE structures compared to their LATTICE or H-FRAME equivalents. Even though monopoles are usually more expensive, they are more common in populated areas where more people will have to see them.

Transmission towers must resist significant loads from wind and the tension of the lines. Their foundations usually consist of drilled CONCRETE PILES deep into the earth. Most towers are designed as suspension structures where the conductors simply hang vertically from the INSULATORS. *Suspension towers* can't withstand much unbalanced force from the conductors. Stronger towers, called *tension towers*, are placed at locations where the line changes direction, crosses a large gap like a river, or requires a block to a cascading collapse that could occur if the conductors were to break. Differentiating between suspension and tension towers is simple: just look at the orientation of the insulators. On suspension towers, they'll be mostly vertical. Any other direction means unbalanced tension in the conductors, requiring a stronger tower.

Lightning represents a major vulnerability for overhead electric lines. A lightning strike can send a massive surge of high voltage down the wires, leading to *arcs* (also called *flashovers*) and damaged equipment. Overhead transmission lines typically include at least one non-energized line running along the tops of the pylons. These are called SHIELD WIRES and are intended to capture lightning strikes so that the main conductors are not affected. Stray voltages are harmlessly routed to ground at each tower. If you look closely, you can often see copper conductors at the bottom of the tower, which are connected to either separate GROUNDING ELECTRODES or the steel reinforcement within the concrete foundation piles. Transmission providers occasionally include a fiber-optic cable within the core of the shield wire for use in their communications network.

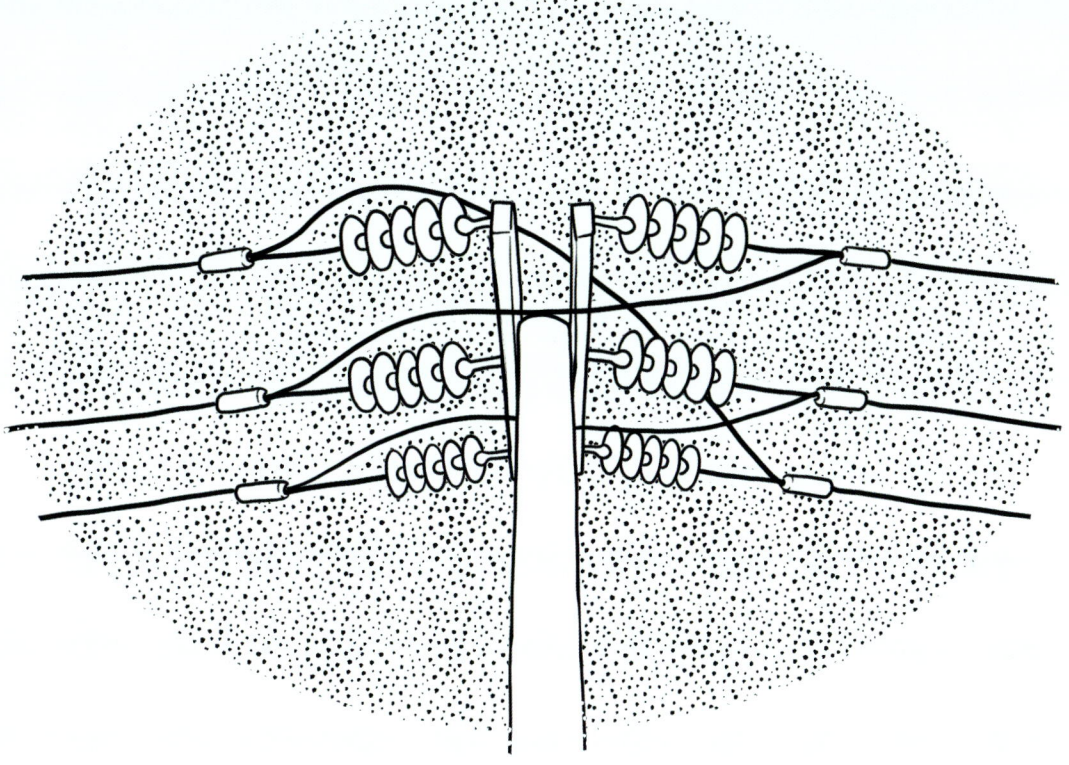

Magnetic fields created by each high-voltage transmission line and the environment can distort the current flowing in parallel conductors. The arrangement of the phases to each other and to the ground means that the flow of electricity in each conductor will be warped in a slightly different way. To balance out the distortion between each of the three phases, long transmission lines need to be "twisted" at regular intervals along the way. Look for special towers called *transposition towers* that allow conductor phases to swap locations before continuing onward.

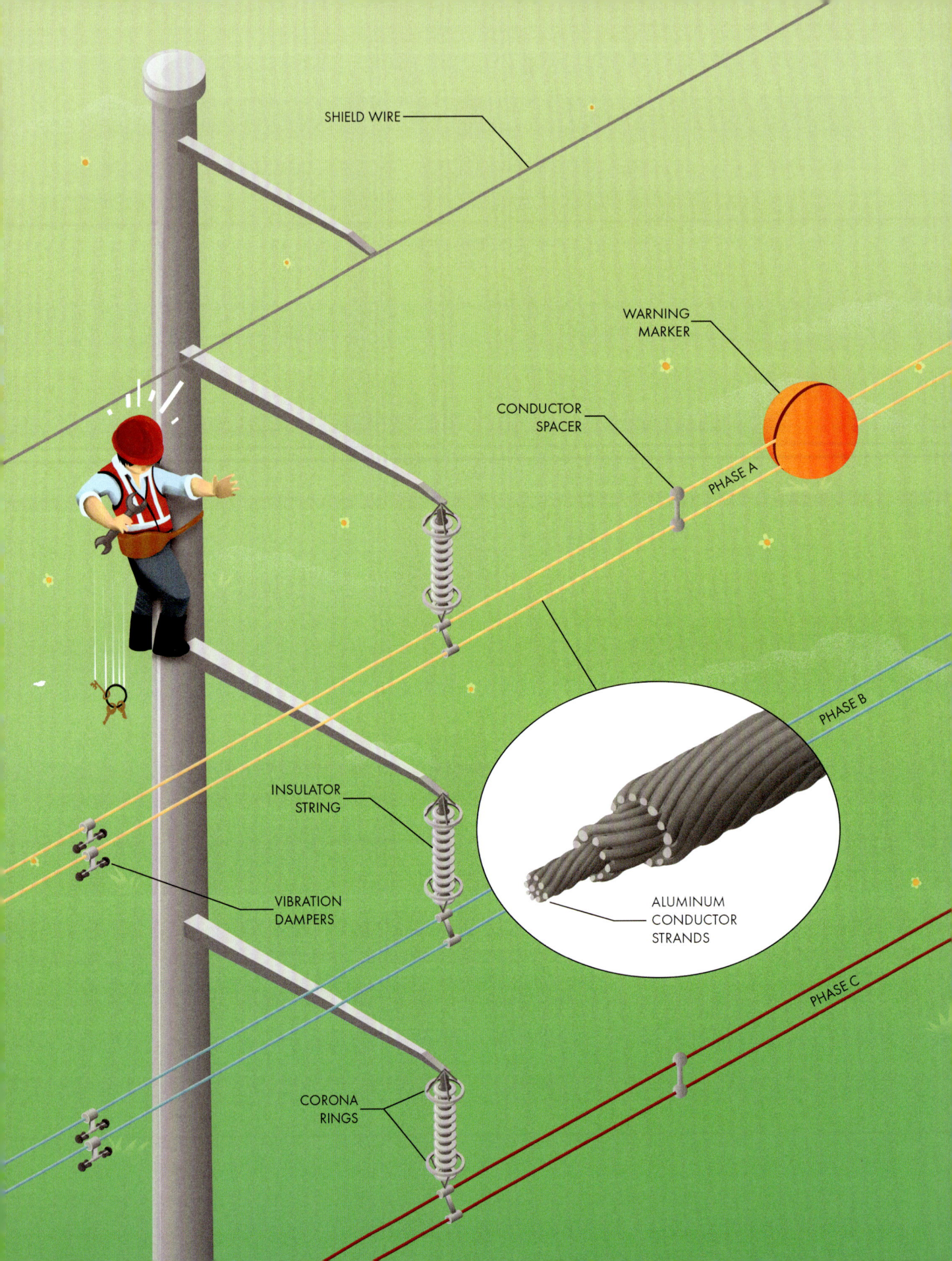

SHIELD WIRE

WARNING
MARKER

CONDUCTOR
SPACER

PHASE A

PHASE B

INSULATOR
STRING

VIBRATION
DAMPERS

ALUMINUM
CONDUCTOR
STRANDS

PHASE C

CORONA
RINGS

Transmission Line Components

Unlike a typical household extension cord, transmission lines are more than just a group of wires. Their enormous scale and high voltages create many engineering challenges to overcome. A variety of equipment and components arise from the need to make transmission lines efficient, cost-effective, and safe (both for the workers who maintain them and the public).

Of course, the most important components are the lines themselves. Conductors are almost always made from many individual STRANDS of aluminum. Aluminum is a great choice because it's lightweight, doesn't easily corrode, and offers low resistance to electrical current. But, if you've ever crushed a soda can, you know that aluminum is not particularly strong compared to other materials. Transmission conductors not only need to carry the electricity, but they must also span great distances between each pylon and withstand the forces of wind and weather. They can also become hot when moving a great deal of electrical current. This heat causes the lines to sag as the metal conductors expand. If they sag too far, conductors can come into contact with tree branches or other obstacles, creating a dangerous short circuit or even starting a fire. For these reasons, aluminum cables are often reinforced with steel or carbon fibers for extra strength.

Another difference compared to a household extension cord is that the conductors of high-voltage transmission lines are bare. They have no outer jacket of insulation. The amount of rubber or plastic that would be required to prevent electrical arcs would add too much weight and cost to the wires. Instead, most of the insulation for high-voltage lines comes from air gaps, simply maintaining large amounts of space between the energized lines and anything that might serve as a path to ground. You might see the challenge here. The conductors can't float in the air without support, but anything they touch becomes dangerously energized. If they were connected directly to the towers, it would create a severe hazard to anyone or anything on the ground (not to mention a short circuit between each phase). So, conductors are instead connected to each tower through long INSULATOR STRINGS.

The design and construction of these insulators are critical because they are the only connection between conductors and towers. Traditionally, insulators have been made from a string of ceramic discs (usually glass or porcelain). The discs lengthen the flow path of electricity leakage if the insulator gets wet or dirty, reducing how much power can escape. These discs are also somewhat standardized in size, so counting them provides an easy way to roughly guess a line's voltage: multiply the number of discs by 15 kilovolts (kV). Nonceramic insulators are becoming more popular, including those made using silicone rubber and reinforced polymers. Unfortunately, the

15 kV per disc rule of thumb doesn't apply for the newer nonceramic insulators, so you'll have to use other clues to guess the voltage of the line.

The high voltages used in transmission lines can lead to some interesting phenomena. For one, the alternating current creates a *skin effect* where most of the current travels around the surface of the conductor rather than evenly through the full area. That means increasing the diameter of a conductor doesn't always create a corresponding increase in its ability to carry electricity. Also, power on the lines can be lost to *corona discharge*, an effect created from ionization of the air surrounding the conductors. Listen closely and you can occasionally hear the corona discharge as a sizzling sound, particularly on dewy mornings, during stormy weather, or in high altitudes where atmospheric pressure is low.

Because of these two phenomena, each phase of a high-voltage transmission circuit is sometimes run as a *bundle* of smaller conductors separated by SPACERS rather than a single large one. The smaller diameter conductors are more efficient at transmitting alternating current since they have more area on the surface where the electricity prefers to travel, and the large overall diameter of the bundle reduces corona discharge. One way to estimate the voltage of a transmission line is to count the number of bundled conductors for each phase. Lines below 220 kV commonly use only one or two conductors, and lines above 500 kV often have three or more. Corona discharge is most prevalent at sharp corners and edges of metal surfaces, such as connections to the insulator strings. On transmission lines with very high voltage or in areas that receive a lot of rainfall, you may see CORONA RINGS attached to insulators. These rings distribute the electric field over a larger area, eliminating sharp corners and edges to reduce corona discharge even further.

Wind can affect the conductors, causing oscillations that lead to damage or failure. Over time, this vibration can fatigue the conductor material or cause abrasion at connections, reducing its lifespan. Replacing conductors is a big, expensive job, so utilities want them to last as long as possible. DAMPERS are often installed to absorb wind energy and reduce long-term damage to the wires. Spiral dampers are used for smaller conductors, and larger lines use suspension dampers, also called *Stockbridge dampers*. Not all wind is unwelcome, though. It also provides a beneficial effect by cooling off the wires. Conductors are often reinforced where they attach to insulators to give this critical element extra strength.

Finally, not all human activities occur on the ground well below these dangerous lines. Spheres called WARNING MARKERS are sometimes attached to lines to make them more visible to people who may be operating tall equipment or up in the air themselves. You'll notice them most often near airports and over waterways.

Above a specific voltage and a certain distance, it becomes economical to use direct current instead of alternating current for electrical transmission lines. Although the equipment to convert from AC to DC and vice versa is quite expensive, *high-voltage DC (HVDC)* lines have many benefits over AC. AC power must "charge up" the line each time the current changes direction, which requires a lot of extra power. HVDC lines aren't affected by this effect (called *capacitance*) and thus can be more efficient. HVDC lines are also used as connections between separate power grids where the alternating currents may not be synchronized. HVDC transmission lines use incredible voltages (up to 1100 kV), but they are still relatively rare, especially in North America. They are instantly recognizable because they use only two conductors—positive and negative, just like a battery—rather than the three phases of typical AC lines.

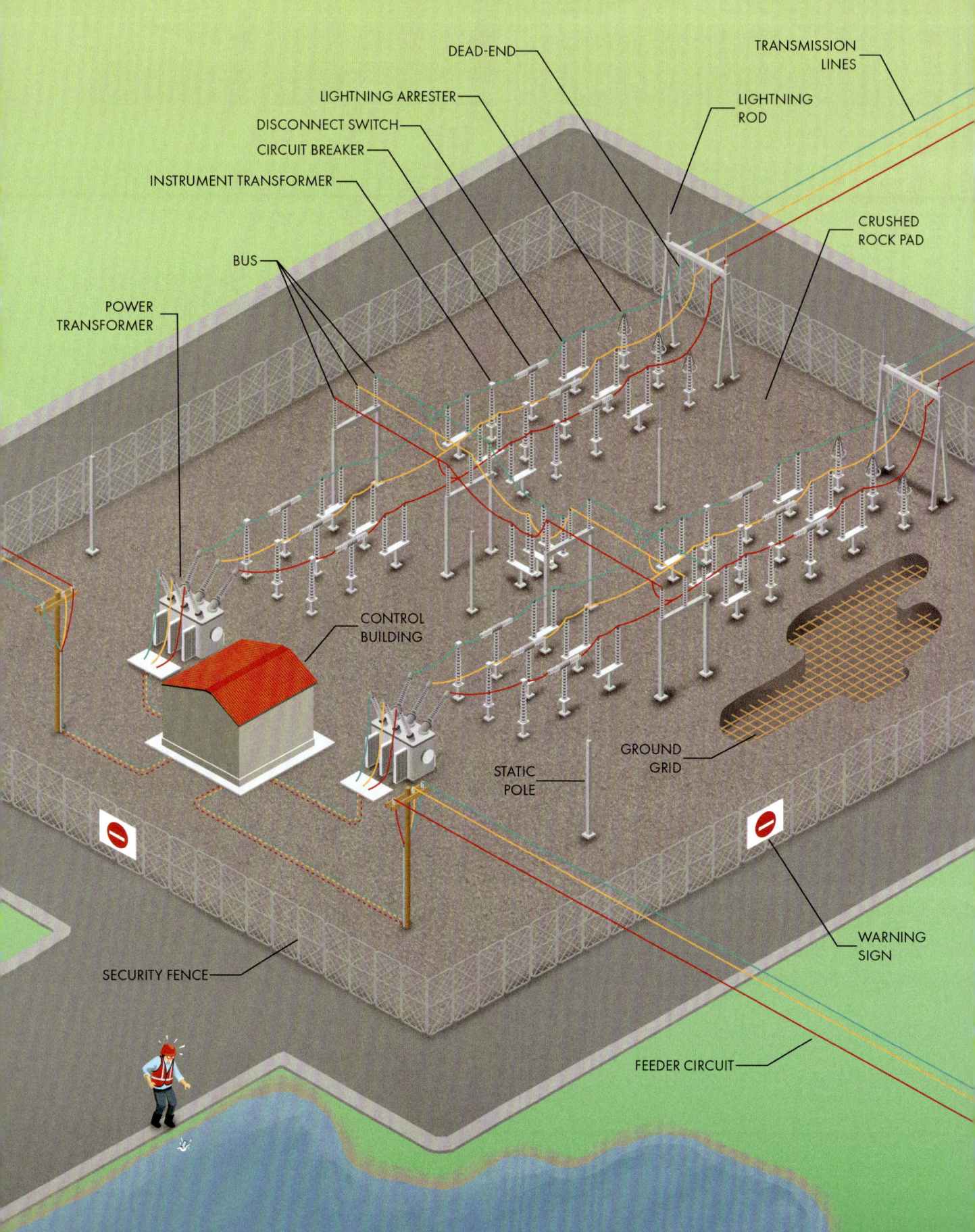

DEAD-END

TRANSMISSION
LINES

LIGHTNING ARRESTER

LIGHTNING
ROD

DISCONNECT SWITCH

CIRCUIT BREAKER

INSTRUMENT TRANSFORMER

CRUSHED
ROCK PAD

BUS

POWER
TRANSFORMER

CONTROL
BUILDING

GROUND
GRID

STATIC
POLE

WARNING
SIGN

SECURITY FENCE

FEEDER CIRCUIT

Substations

If you consider the power grid a gigantic machine, substations are the links that connect the various components together. Originally named for smaller power plants, *substation* has become a general term for a facility that can serve a wide variety of critical roles on the power grid. These roles include monitoring the grid's performance to make sure nothing is awry, changing between different voltage levels, and providing protection against faults. The most commonly seen substation around cities are *step-down* facilities that convert high-voltage transmission to a lower, safer voltage for distribution within populated areas.

At first glance (and sometimes even after a good long stare), substations are a complex assemblage of wires and equipment. When I was a kid, I thought they were playgrounds (to the delight and horror of my parents). For a power-grid greenhorn, mentally untangling these mazes of modern electrical engineering can be challenging, especially because the scaffolding and support structures look so similar to the conductors and bus work. The simplest way to identify energized lines and equipment is to look for which parts are held by insulators. Eventually, you'll be able to follow the current's path as it makes its way through. Each phase of the conductors is highlighted in the illustration to help you trace the paths of flow. (The next section describes specific pieces of equipment in

a substation and their functions in further detail.)

Substations often serve as the termination points of many TRANSMISSION LINES. High-voltage lines enter the substation through a support structure called a DEAD-END that provides support and spacing. These are the only locations where very high-voltage lines drop from their safe heights down to ground level, so extra precaution is required to keep the lines contained.

The heart of a substation and the primary connection between all the various devices and equipment in a substation is the BUS, a set of three parallel conductors (one for each phase). The bus is usually made from rigid overhead tubes running along the entire substation. The substation's overall reliability depends on the arrangement of the bus because different schemes offer different amounts of redundancy. In the event of an equipment failure or regularly scheduled maintenance, utilities don't want to shut down the entire facility, so the bus is designed to reroute the power around equipment that's out of service when necessary.

Substations have a high-voltage side and a low-voltage side, separated by the POWER TRANSFORMERS (discussed in the next section). At step-down facilities, power leaves the substation as individual circuits called FEEDERS. Each feeder has its own circuit breaker, allowing smaller groups of customers to be isolated from the grid in the event of a fault. Many feeders leave the substation underground and resurface at a nearby utility pole for distribution to customers.

Most substation equipment is located outdoors in the open air. However, certain components are more vulnerable to weather and changes in temperature, including relays, operating equipment, and some CIRCUIT BREAKERS. These more sensitive pieces of equipment are often located within a CONTROL BUILDING at the substation. As with transmission lines, lightning poses a severe threat to substations. STATIC POLES and LIGHTNING RODS poke into the air to capture strikes and shunt them directly to ground, protecting costly equipment from surges. ARRESTERS also help deal with the damaging effects of lightning. These devices are connected to energized lines, but they don't normally conduct any current. Arresters instantly become conductors only when they sense a large spike in voltage, safely diverting excess electricity into the earth.

Many substation features observable from the outside are related to safety of the workers who operate and maintain the equipment. One of the most critical factors in protecting equipment and workers in a substation is ensuring that stray electricity has somewhere to go. All substations are built with a GROUND GRID, a series of interconnected copper wires buried below the surface. In the event of a fault or short

circuit, the substation needs to be able to sink lots of current into the ground through this grid to trip the breakers as quickly as possible. This grounding grid also makes sure that the entire substation and all its equipment are kept at the same voltage level, called an *equipotential*. Electricity flows only between points of different voltage potential, so keeping everything at the same level ensures that touching any piece of equipment doesn't create a flow of electricity through a person. The cases and support structures of every piece of equipment are bonded together via the grounding grid.

You might notice that most substations have a layer of CRUSHED ROCK as a floor. This isn't just because lineworkers don't like to mow the grass! Crushed rock is freely draining and doesn't hold moisture, so it provides a layer of insulation above the soil and prevents formation of puddles from rain.

Keeping away from high-voltage facilities is common sense for most people, but as crazy as it sounds, substations are common targets for thieves wanting to steal copper wire. Substations are surrounded by FENCES and WARNING SIGNS to make sure that any wayward citizens know to stay out. If you look closely, you'll notice that even the fences have wires connecting them to the subsurface grounding grid ensuring the equipotential extends not only to workers inside the fence, but also to anyone on the outside.

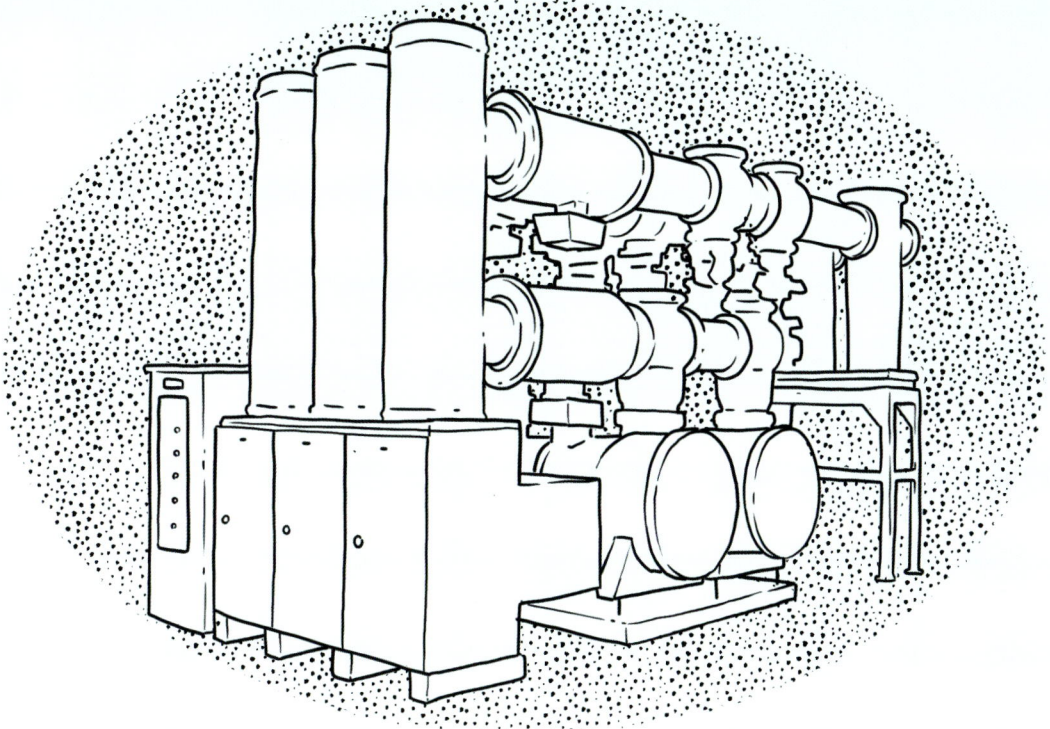

Much of the equipment used in outdoor substations is called *air insulated switchgear* because it uses ambient air and spacing to prevent high-voltage arcs from forming between energized components. Another type of equipment called *gas insulated switchgear* involves encapsulated equipment in metal enclosures filled with a dense gas called *sulfur hexafluoride*, which allows installation of high-voltage components in locations where space is limited. You'll have to be lucky to see a substation consisting entirely of gas insulated switchgear because they are much costlier and, thus, rare. Gas insulated switchgear is also more likely to be hidden inside a building and protected from weather, rather than exposed to the open air. You'll know you've spotted one from the characteristic tight clusters of metal piping, lots of bolted flanges, and many components in groups of three to handle each phase of power.

POWER TRANSFORMER

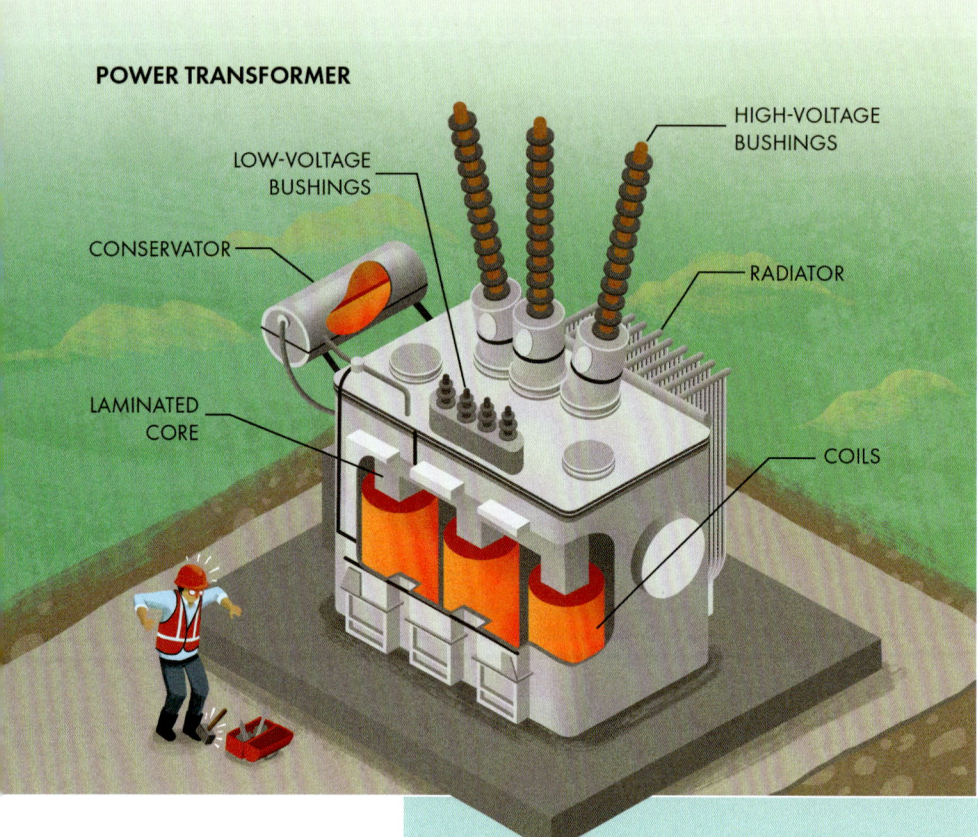

LOW-VOLTAGE
BUSHINGS

HIGH-VOLTAGE
BUSHINGS

CONSERVATOR

RADIATOR

LAMINATED
CORE

COILS

INSTRUMENT TRANSFORMERS

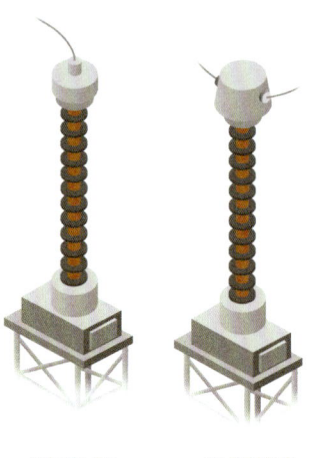

VOLTAGE
TRANSFORMER

CURRENT
TRANSFORMER

DISCONNECT SWITCHES

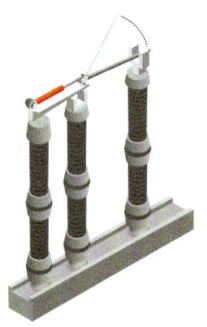

HINGED DISCONNECT

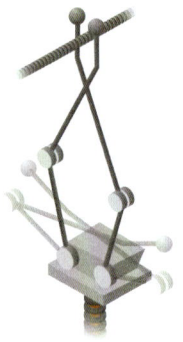

PANTOGRAPH
DISCONNECT

CIRCUIT BREAKERS

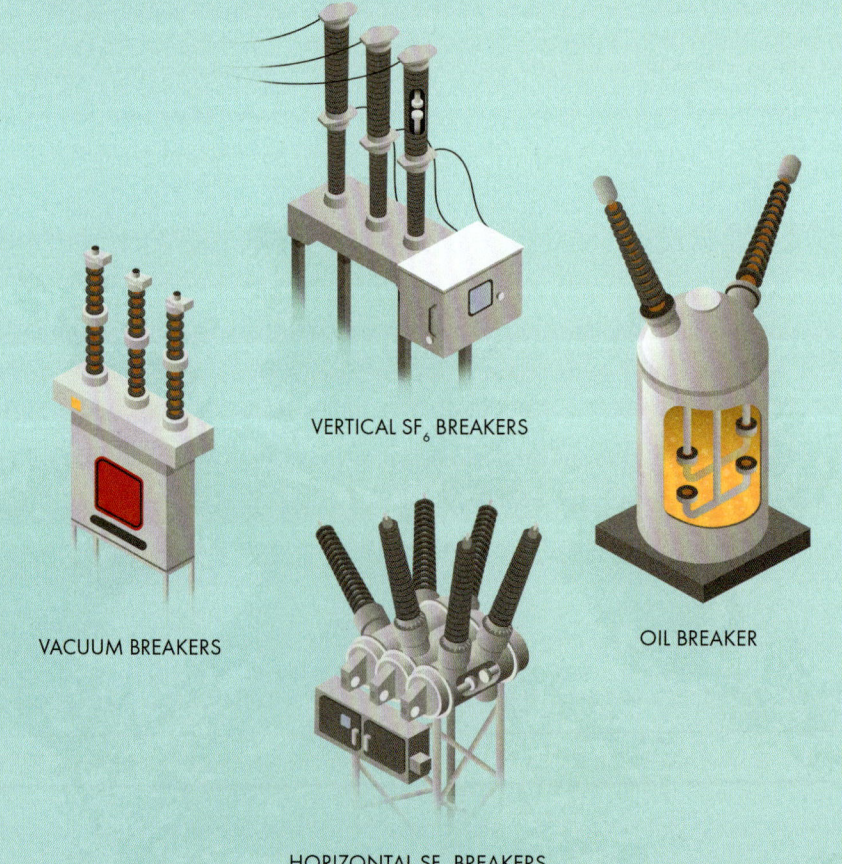

VERTICAL SF$_6$ BREAKERS

VACUUM BREAKERS

OIL BREAKER

HORIZONTAL SF$_6$ BREAKERS

Substation Equipment

Understanding the layout and flow of electricity in a substation is only half the story. Substations are made up of many different individual pieces of equipment, each of which serves an important role. The joy of substation spotting is made much greater by being able to identify those pieces of equipment and understanding how they work.

One of the most important jobs at a substation is stepping up or down voltage; that is, converting between the more efficient (but more dangerous) high voltage from transmission lines and the lower (and easier to insulate, although still quite dangerous) voltage for the smaller lines within urban areas. This conversion is done using a POWER TRANSFORMER, a device that relies on the alternating current of the grid to function with no moving parts by taking advantage of *electromagnetism*. A transformer mainly consists of two adjacent COILS of wire. The alternating current of the input electricity generates magnetic fields that are focused and directed by a LAMINATED CORE consisting of many thin sheets of iron. These magnetic fields couple to the adjacent coil, inducing a voltage in the output wires. The voltage out of the transformer is proportional to the number of loops in each coil. Transformers are usually the largest and most expensive pieces of equipment in the entire substation, so they are easy to identify.

The insulators guiding conductors into and out of the transformer are called BUSHINGS. They support the energized lines as they pass through the metal case into the transformer, protecting against short circuits. You can easily tell which lines are higher- and lower-voltage by the difference in the size of the bushings. The higher the voltage, the larger the bushings need to be to maintain enough distance to prevent arcs.

Although grid-scale transformers are very efficient, they still lose some power to noise and heat. If you get close enough, you'll definitely notice the low-pitched hum that occurs because the constantly changing magnetic fields cause vibrations to the components inside the transformer. Heat is also generated from the resistance in the copper coils and can eventually damage the transformer. Transformers are usually filled with oil to help with cooling. RADIATORS consisting of fans and heatsinks can be seen on the outer metal case to dissipate heat and help keep the oil and components cool. You may even see a smaller tank (called a CONSERVATOR) on top of a transformer case to hold extra oil and allow the fluid to expand and contract.

Nearly every line and piece of equipment in a substation needs to be isolated completely from the rest of the energized system during maintenance or repairs. DISCONNECT SWITCHES are usually installed on each side of the equipment for this reason. They can't interrupt large currents through the system and are used strictly for isolating equipment to keep workers safe. The most common disconnect switches are motor-operated and consist of a HINGED blade and a stationary contact, both of which are mounted on insulators. PANTOGRAPH disconnect switches raise and lower with a scissoring action to connect to bus bars.

On occasion, it is necessary to interrupt the flow of electricity on some part of the power grid. Most commonly, interruption is needed due to a fault, which can cause significant damage to costly and vital equipment. CIRCUIT BREAKERS provide the means to stop the flow of electricity, allowing faults to be isolated from the rest of the system. They not only protect the other equipment on the grid, but also make problems easy to find and repair quickly. Interrupting current on energized lines isn't as simple as it sounds, though. Just about anything can conduct electricity if the voltage is high enough, and that includes air. Even if you create a break in the line to disconnect it, electricity can continue flowing through the air in a phenomenon known as an *arc*. Arcs need to be extinguished as quickly as possible to prevent damage to the breaker or unsafe conditions for workers, which means all circuit breakers for high-voltage equipment need to include some type of arc suppression.

For lower voltages, the circuit breakers are located in a sealed container under VACUUM to avoid electricity conducting in the air between the contacts. For higher voltage, breakers are often submerged in tanks filled with nonconductive OIL or dense gas called SULFUR HEXAFLUORIDE (SF_6). Another option is to use a massive blast of air to blow out the arc. All breakers are connected to devices called *relays* that can automatically trigger during a fault condition. Breakers can also be manually operated to remove a circuit from service as needed for maintenance or to shed load during periods of extreme electrical demand. Because many faults are temporary (for example, lightning strikes), some breakers, called *reclosers*, automatically reenergize the circuit if the fault has cleared.

Relays monitor the voltage, current, frequency, and other parameters on the grid to identify problems and trigger the breakers, but we can't just feed high voltage into sensitive operating equipment. Instead, special transformers called INSTRUMENT TRANSFORMERS convert the high voltages and currents on the conductors to smaller, safer levels that can be sent to the relays. Instrument transformers are the eyes of the power grid, monitoring conditions to make sure everything is working properly. Although they look similar, there's an easy way to tell them apart: the primary coil for VOLTAGE TRANSFORMERS is usually connected between one phase and ground, so you'll only see one high-voltage terminal. The primary coil for a CURRENT TRANSFORMER is connected inline (that is, in series) with the conductor, so there will be two high-voltage terminals.

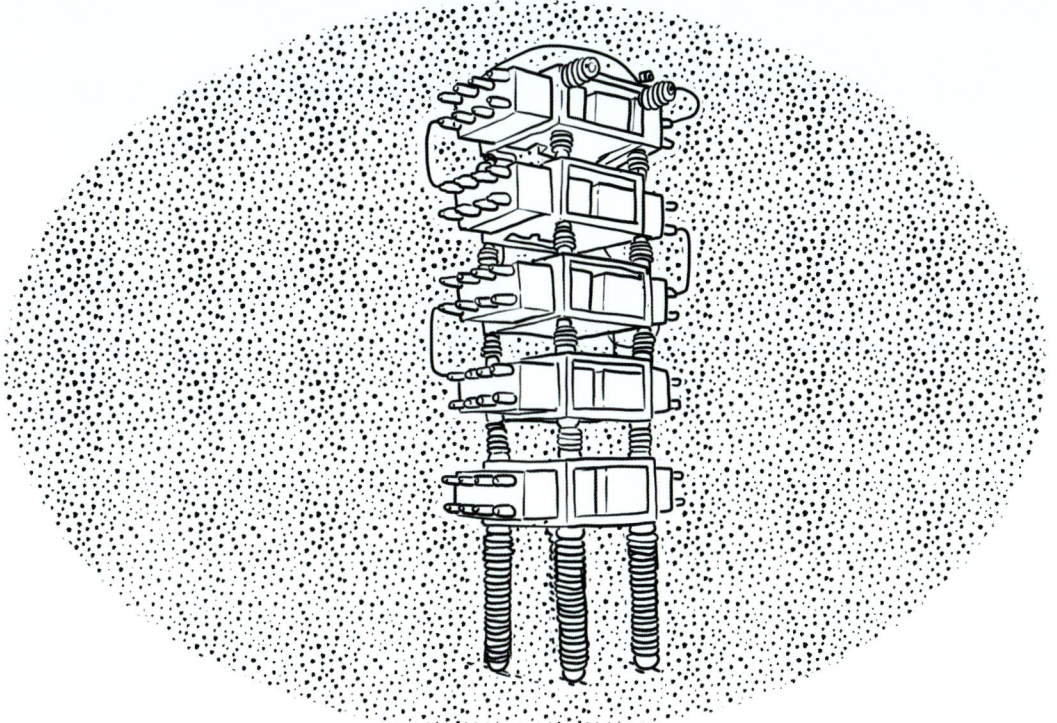

One challenge with AC power is that the voltage and current can lose synchronization. Certain kinds of electrical loads are reactive, meaning they momentarily store power before returning it to the grid. This causes the current to lag or get ahead of the voltage, reducing its ability to perform work. It also reduces the efficiency of all the conductors and equipment powering the grid because more electricity has to be supplied than is actually being used. The measure of this reduction is called the *power factor*. Some substations include banks of capacitors to bring the current and voltage back into sync and help improve the power factor in the lines. The capacitors absorb some or all of the mismatch in voltage and current, allowing more efficient use of conductors, transformers, and other equipment and helping stabilize the voltage on the grid. Look for arrays of small boxes on steel racks.

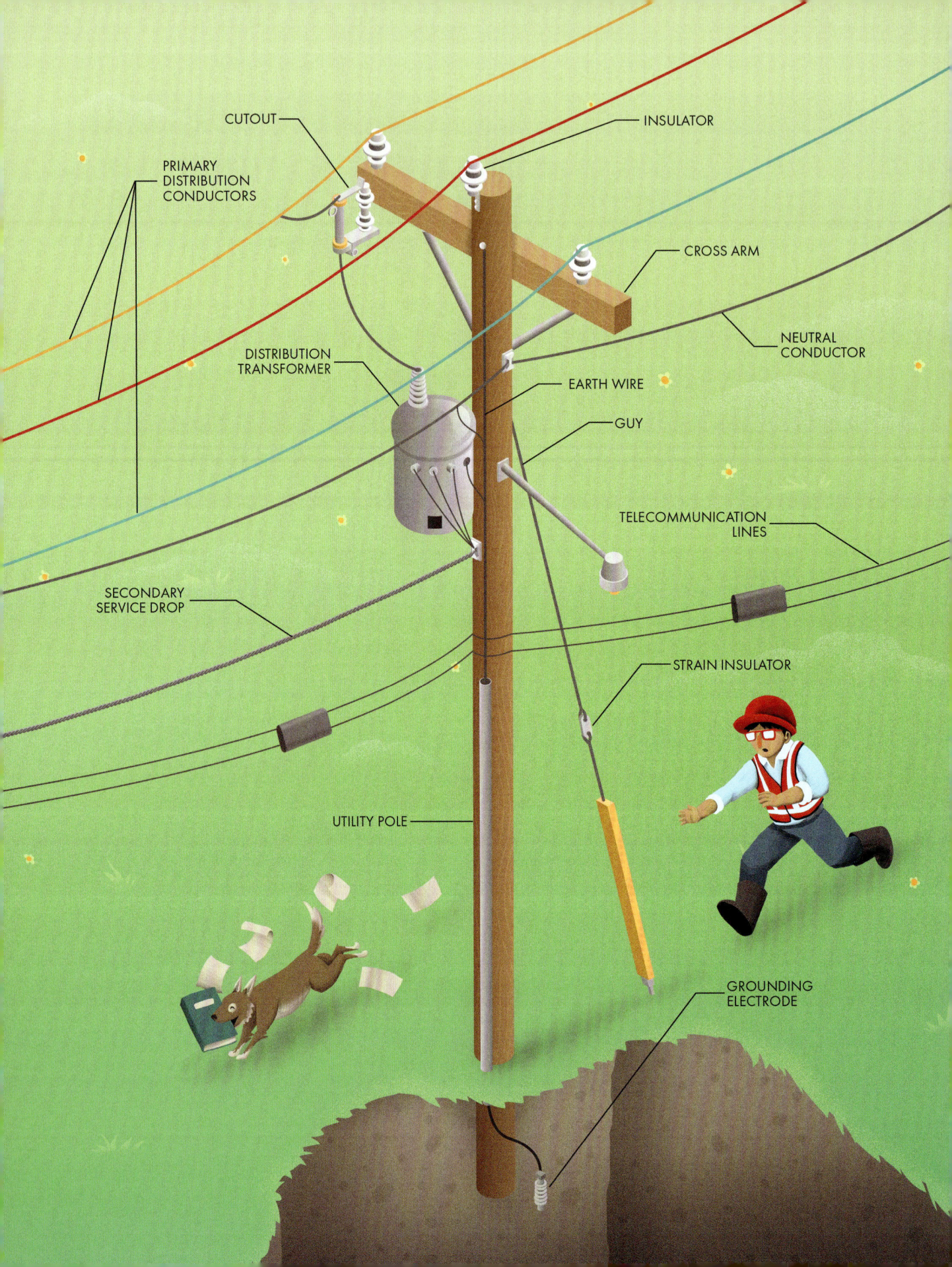

CUTOUT

INSULATOR

PRIMARY
DISTRIBUTION
CONDUCTORS

CROSS ARM

NEUTRAL
CONDUCTOR

DISTRIBUTION
TRANSFORMER

EARTH WIRE

GUY

TELECOMMUNICATION
LINES

SECONDARY
SERVICE DROP

STRAIN INSULATOR

UTILITY POLE

GROUNDING
ELECTRODE

Typical Utility Pole

Almost nothing is more ubiquitous in the constructed world than the UTILITY POLE, which serves a critical role in the distribution of electricity on the grid. *Distribution* describes the portion of the power grid that brings electricity to all the individual consumers. If transmission lines are electrical highways, distribution lines are the residential streets. They usually start at a substation where individual power lines (called feeders) fan out to connect to residential, commercial, and industrial customers. In some ways, distribution is nearly identical to high-voltage transmission. Wires are wires, after all. But in other ways, it is surprisingly different. The most obvious difference is that the voltages come down to levels that are easier to insulate, so the heights of the poles and conductors get lower as well.

In most parts of North America, wood is a relatively abundant resource, so it is the material that makes up the vast majority of utility poles. *Preservatives* are used to treat the wood and slow down deterioration from weather and insects. Standards vary regionally, but poles of normal height are usually buried 2 to 3 meters (6 to 9 feet) into the earth. Most utility poles have their own EARTH WIRE running down the pole attached to an ELECTRODE driven into the ground. This wire provides a safe path for any stray currents instead of allowing them to travel through the pole itself, which could lead to shocks or fires.

Poles in a straight line need to support only the vertical weight of the wires atop, but if a pole serves as a corner or dead-end, it experiences a pull to one side. Even if this tension isn't substantial, the long pole acts like a lever, magnifying the force into the ground and potentially toppling it altogether. Whenever the horizontal forces on a pole aren't balanced, GUYS are used for additional support. Each guy is equipped with a STRAIN INSULATOR to make sure that, in the event of an accident, dangerous voltages can't reach the lower section of the cable.

The PRIMARY DISTRIBUTION CONDUCTORS (or *lines*) you see at the top of utility poles are considered medium voltage and usually range from 4 kV to 25 kV. Energized lines are easy to identify because they are supported by INSULATORS. Even though they are at a much lower voltage than transmission lines, the voltage of primary distribution lines is still too hazardous for use in homes and businesses. DISTRIBUTION TRANSFORMERS (described more in the next section) reduce the voltage to its final level—often called *mains* or *secondary voltage*—for use by regular customers. The SECONDARY SERVICE DROPS that connect each customer to the grid are located below the primary conductors. For the safety of workers, the energized lines are always at the top of the poles with space to work between them and other TELECOMMUNICATION LINES (such as cable, telephone, and fiber-optics). See Chapter 2 for more information on communications infrastructure, which often runs parallel to distribution lines on utility poles.

One major difference from transmission lines is that the number of conductors on the distribution grid increases from three to four. This is due to how electrical demands are distributed between each of the grid's three phases. All electrical circuits are loops, so they require two lines: one to supply the current and one to return it to the source. On high-voltage transmission lines, the electricity usage between each of the three phases is perfectly balanced, eliminating the need for a separate return path for electricity. Each pair of phases serves as a source and a return path at the same time. However, on the distribution side, it's not always so simple. Many electrical consumers (including most residential homes) make use of only a single phase. In fact, on the distribution grid, the three phases are often split from one another to service different areas entirely. Look around some residential neighborhoods and you may see many poles with only a single primary conductor and no CROSS ARM. Grid operators try to arrange distribution lines to make sure that all the loads on each phase will be roughly equal, but they are never perfectly in sync. These imbalances between phases necessitate a NEUTRAL CONDUCTOR to act as a return path for stray current.

Much of the complexity of the power grid is due to how we protect it when things go wrong. The grid got its name for a reason. It's an interconnected system, which means that, if we're not careful, small problems can sometimes ripple out and impact much larger areas. Engineers establish zones of protection around each major piece of the power grid using fuses and circuit breakers to isolate faults and make them easy to find and repair. These devices create "managed failures" where you have some loss of service at the cost of protecting the rest of the system (just like the breakers in your house). The goal is that isolating equipment when things go wrong speeds up the process and reduces the cost of making repairs to get customers back online. When your power goes out, it's easy to be frustrated at the inconvenience, but consider also being thankful that it probably means things are working as designed to protect the grid as a whole and ensure a speedy and cost-effective repair to the fault.

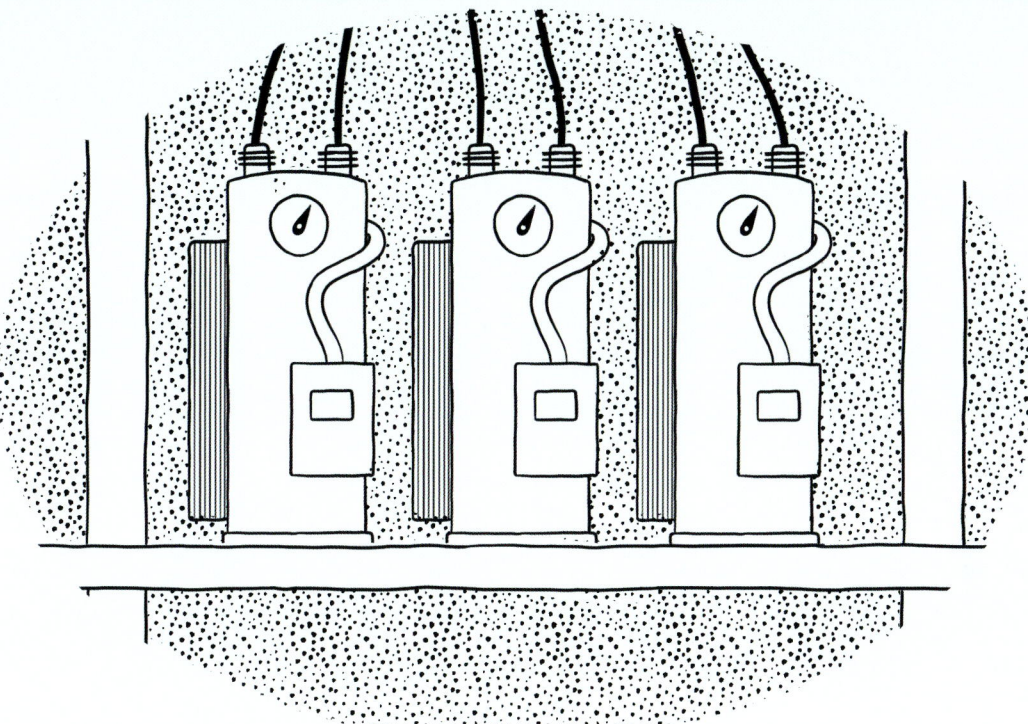

Rural areas often have long runs of primary distribution lines. These long distances create extra resistance and make maintaining a steady voltage level difficult. Another challenge is the increasing popularity of grid-connected solar panel installations. Clouds temporarily casting shadows on the panels can create instability in the distribution voltage for areas with lots of connected panels. *Voltage regulators* are devices with multiple taps that can make small adjustments to the distribution voltage. They work similarly to transformers but make only minor adjustments to voltage, usually plus or minus 10 percent. Regulators monitor voltage on the line directly or perform automatic calculations of voltage drop based on the measured current to adjust their taps up or down. They also look like distribution transformers with cylindrical casings (one per phase). However, they have a few recognizable differences. Both the input and output of a regulator are attached to the primary distribution line, and both bushings are the same size. Also, look for the dial at the top of the canister, which indicates the regulator's tap position. If you're lucky, you might catch it automatically switching positions to maintain the correct voltage on the line.

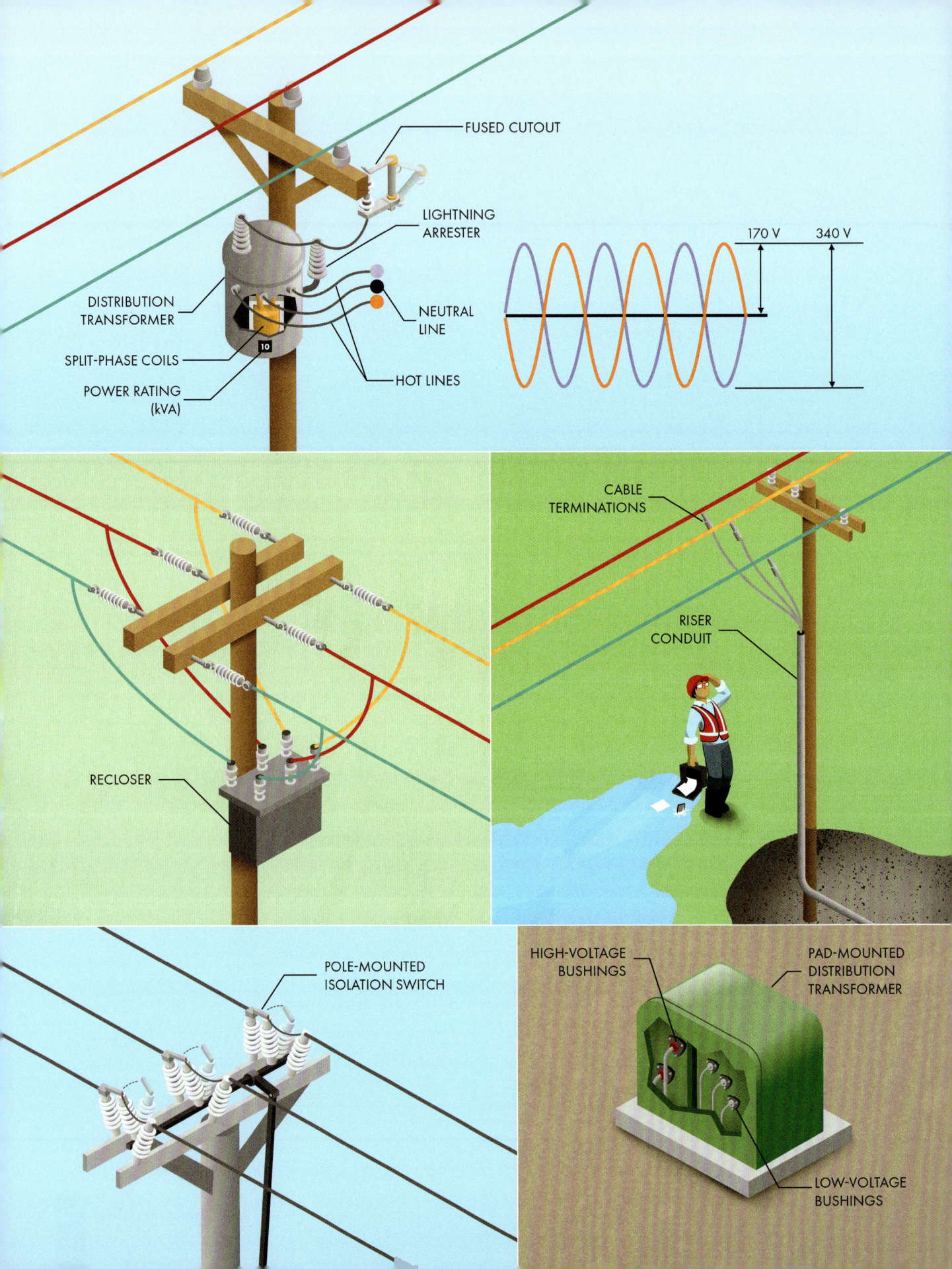

FUSED CUTOUT

LIGHTNING ARRESTER

DISTRIBUTION TRANSFORMER

SPLIT-PHASE COILS

POWER RATING (kVA)

NEUTRAL LINE

HOT LINES

10

170 V   340 V

CABLE TERMINATIONS

RISER CONDUIT

RECLOSER

POLE-MOUNTED ISOLATION SWITCH

HIGH-VOLTAGE BUSHINGS

PAD-MOUNTED DISTRIBUTION TRANSFORMER

LOW-VOLTAGE BUSHINGS

Electrical Distribution Equipment

Like all the other parts of the grid, distribution of electricity requires various pieces of equipment to help with reliability and safety. Just like in a substation, one of the most important pieces of equipment on the distribution grid serves to change the voltage. Although significantly lower than transmission voltages, primary distribution circuits are operated at many thousands of volts, still much higher than can be safely used in most houses and businesses. In most cases, another transformer (called a DISTRIBUTION TRANSFORMER) is needed to step down the voltage to the level generally used in buildings by lights, appliances, and other devices. These transformers often appear as gray canisters just below the conductors on utility poles. They are filled with oil, just like the transformers at substations, and work in almost the same way.

One interesting difference in many places around the world is that the output of distribution transformer coils uses a SPLIT-PHASE design. In this configuration, two energized (or HOT) lines are supplied to the customer with one NEUTRAL conductor connected to ground. One of the energized lines is inverted from the other. In this way, smaller appliances can use the line-to-neutral voltage, about 120 V nominal (170 V peak-to-peak) in most of North America. Devices requiring more power (such as heaters, air conditioners, and clothes dryers) can be connected between the two energized lines, receiving double the voltage. In residential settings, a single distribution transformer can often supply multiple homes. Take a look outside your house and you may notice you share a transformer with a few of your neighbors. Customers with bigger equipment (for example, large air conditioning units) can take advantage of all three phases on the grid. In this case, you may see three single-phase transformers grouped together on the same pole. Look for the POWER RATING on the side of the transformer in *kilo-Volt-Amperes* (*kVA*, roughly equivalent to *kilowatts*).

Just like transmission lines and substation equipment, the distribution grid needs protection from faults and lightning strikes. Much of the hardware you see on pole tops is for when things go wrong. One common protective device is the FUSED CUTOUT, which serves as both a circuit breaker and an isolation switch. The fuse automatically protects a service transformer from short circuits and voltage surges. If the current in the fuse gets too high, the element inside melts, breaking the circuit and disengaging a latch, allowing the fuse door to swing down. These fuses often include an explosive liner to help extinguish the arc that forms inside, so you might hear a loud pop if one trips nearby. It is often so loud that many people assume the transformer has exploded, when really it was protected from damage by the fuse.

Even if the fuse in a cutout hasn't blown, lineworkers can disengage it to isolate the line for maintenance or repair. However, fuses are the simplest protective devices.

More sophisticated circuit breakers can occasionally be seen, including RECLOSERS, which are usually housed in small cylindrical or rectangular canisters. Reclosers open when a fault is detected, then close again to test whether the fault has cleared. Most faults on the grid are temporary, such as lightning or small tree limbs making contact with energized lines. Reclosers protect transformers without requiring a worker to come replace a fuse for minor issues. They usually trip and reclose a few times before deciding that a fault is permanent and locking out. If you ever lose and regain electricity in a short period of time, a recloser is probably why. Other types of ISOLATION SWITCHES atop utility poles help lineworkers perform maintenance or make repairs. Many use a mechanism to disconnect all three phases at once. Finally, like other parts of the grid, distribution lines use ARRESTERS to redirect surges in voltage from lightning strikes safely to ground.

Not all distribution of grid power happens overhead. In the urban core of many cities, you'll hardly see any overhead lines at all. Instead, power is run in ducts below the ground. Also, newer residential and commercial developments often elect to bury distribution lines to avoid the untidy and cluttered appearance of overhead wires. Using underground distribution lines is not a trivial choice, since they are far more expensive to install and often require more time to repair when damaged. However, these lines are better protected from weather and don't impose on the aesthetics of the urban landscape. Even if not run continuously underground, it is not uncommon for a distribution line to dive belowground and come back up shortly after to avoid an overhead hazard or even to keep a sign from being obstructed.

Although you can't see underground distribution lines, you can often see where they start and stop. Look for utility poles with large RISER CONDUITS attached. Underground power lines must have an insulating jacket to protect them from moisture and short circuits. Insulation around conductors can't just start and stop wherever because moisture could get inside from the ends. CABLE TERMINATIONS (colloquially called *potheads*) are used to seal the transition between insulated and bare cables.

Another location where underground wires come up to the surface is at a transformer. Although less visually intrusive than its overhead equivalent, the PAD-MOUNTED DISTRIBUTION TRANSFORMER serves as a reminder that the power grid still exists in areas without overhead lines. You may be curious what's inside those green cabinets. It's the exact same device as in the ones mounted overhead. The cabinet door provides access to the high- and low-voltage BUSHINGS just like you see on a pole-mounted transformer.

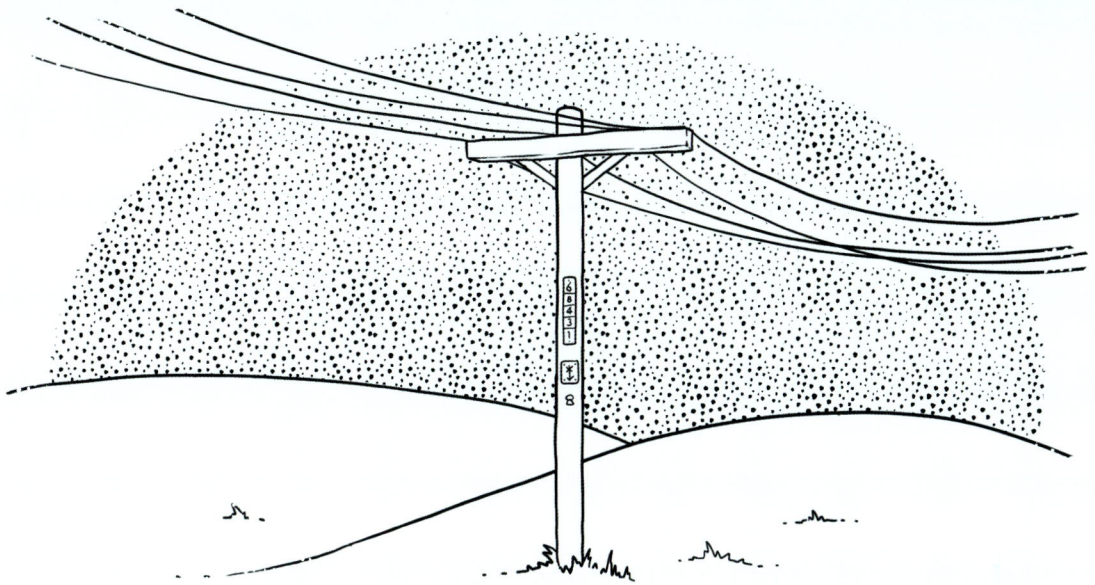

Utility poles are often adorned with cryptic markings and metal tags. Sometimes they are simply identifiers for the utility or a manufacturer's mark, but not always. Red tags with arrows are warnings to lineworkers that the pole is damaged and to be careful or avoid climbing the pole altogether. Pole tags can also indicate the last time the pole was inspected and what type of treatments have been applied to protect it from bugs and rot. Finally, stamps in the wood offer clues about where the pole was manufactured, the wood species used, and even the length of the pole. Keep an eye out for different kinds of markers and see if you can decipher their meaning.

2

COMMUNICATIONS

Introduction

Communication isn't unique to the human species, but telecommunication is. Sharing information beyond a shouting distance requires plenty of innovation. Many of humankind's most significant developments have been ways of sending and receiving messages across expanses. From smoke signals and carrier pigeons to GPS and the internet, telecommunications have profoundly shaped the ways we live, work, and play.

This chapter explores how we send and receive information over long distances, and, most important, the infrastructure that makes it all possible—or at least it does at the time of writing. No other part of society seems to be changing more rapidly than our communications technologies. In 10 years, this chapter may seem outdated. In 20 years, the technology described here may be unrecognizable. It is already easy to take such systems for granted in the Information Age, but captivating details still exist in the engineering behind how we transfer and share knowledge, entertainment, and so much more.

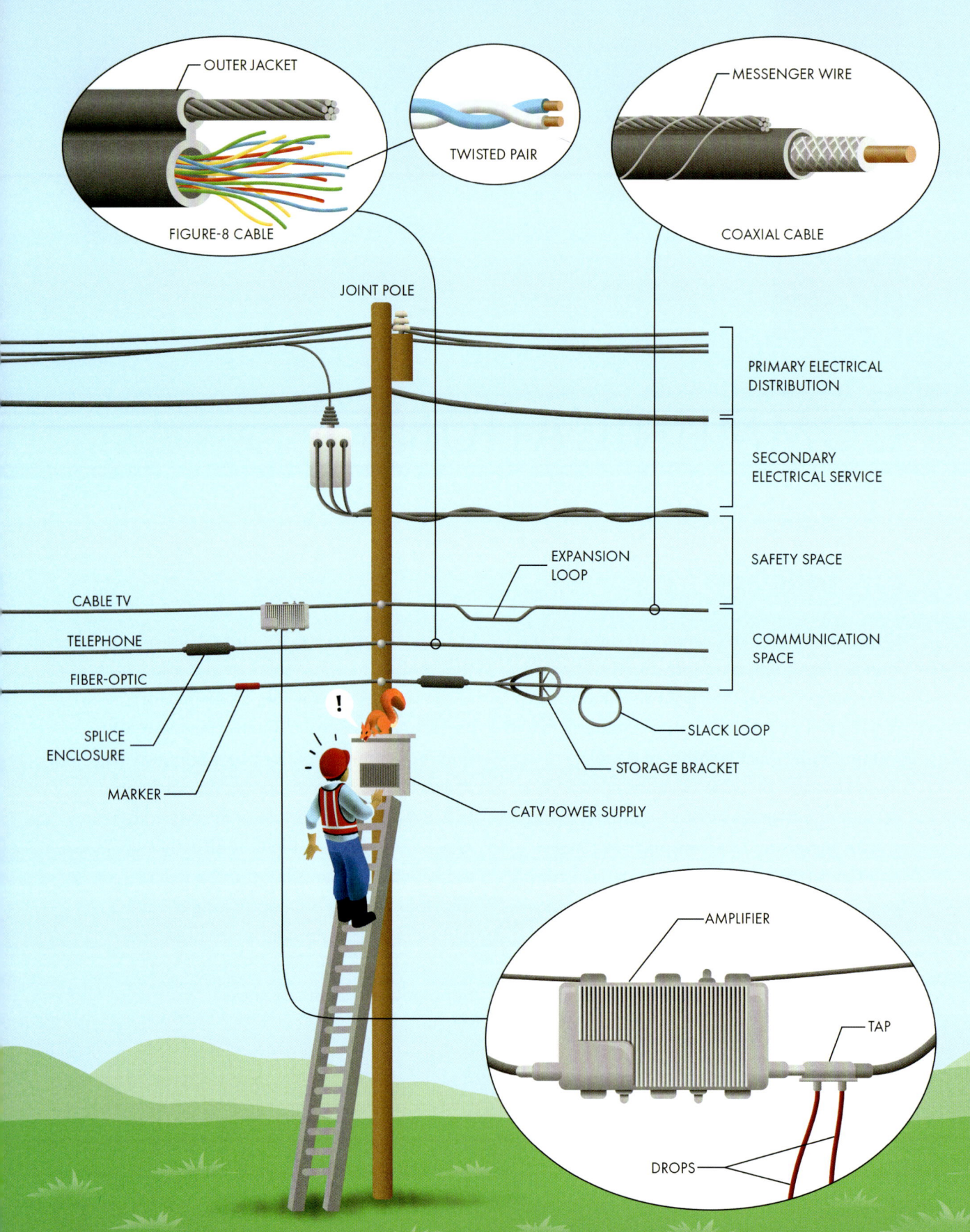

Overhead Telecommunications

The majority of our *telecommunications* occurs through physical lines, either metal wires or glass fibers, and these lines can run in essentially two places to avoid conflicts with other human activities: through the air or below the ground (underwater is a third option in some situations). This section covers the former, and the next section discusses underground installations.

Overhead communication lines are almost always carried on poles along with other utilities. Chapter 1 provides a look at utility poles for electricity distribution, but that's not their only use. JOINT POLES are those shared among multiple utilities. Not every joint pole supports every kind of utility, but no matter which lines run along the pole, all their locations are carefully prescribed. PRIMARY ELECTRICAL DISTRIBUTION lines run along the tops of the poles, farthest from the ground since they have the greatest potential for danger. The SECONDARY ELECTRICAL LINES that service customers run directly below them. Between the electrical and communications lines is a SAFETY SPACE reserved for utility workers to make connections and perform maintenance without being exposed to the danger of high-voltage lines. The COMMUNICATION SPACE is the lowest along the poles, because these lines don't pose a shock hazard and require more frequent maintenance.

Although many different types of communication lines can be strung overhead along poles, only three make up the preponderance of what you can observe on a standard pole: TELEPHONE, coaxial CABLE TV, and FIBER-OPTIC lines. Seeing all three running in parallel on the same poles isn't uncommon, and telling them apart is easy if you know what to look for.

Stringing wires across long distances can create significant tension forces, and most communication lines aren't meant to support their own weight as they span from pole to pole. Instead, a steel MESSENGER WIRE provides this needed support. The communication cables are lashed to the messenger wire, or in the case of FIGURE-8 CABLES, the messenger wire is incorporated into the protective OUTER JACKET.

Although the network of copper lines that make up the *plain-old telephone service (POTS)* are quickly being phased out, they still can be seen on poles around the world. Since 1876, we've been transmitting voice signals along dedicated circuits of copper wire, and it's still the simplest way for a home or business to connect to the telephone network in many places. Each landline consists of a TWISTED PAIR of thin copper wires. Since every household and business can have its own direct lines to the *local telephone exchange*, the cables can grow quite large, sometimes containing hundreds or thousands of pairs. The lines join together into larger and larger cables at splices, easily visible by the boxy black SPLICE ENCLOSURES seen near poles.

All these wires running in parallel alongside each other would naturally create electromagnetic interference and "cross-talk" between circuits. However, twisting each pair of wires in a phone line creatively solves this problem because the undesired interference affects each wire of the twisted pair equally. The desired communications signal is sent in the voltage difference between the two wires of the twisted pair, so any unwanted voltage common to both wires is subtracted out.

Another ubiquitous telecommunication medium is the cable TV network (often abbreviated as *CATV*). Despite the name, most CATV networks support telephone and high-speed internet service in addition to television programming. Like POTS, the CATV network begins at a central location, which is called the *head-end*. From there, the signals are distributed mostly using COAXIAL CABLE, named because the inner conductor and surrounding metal shield are concentric around a common axis. These cables can carry high-frequency radio signals with very low losses or issues with interference because of the shielding effect of the outer conductor. They start as large trunks that feed multiple distribution lines. AMPLIFIERS (also called *line extenders* and recognizable by their heat-dissipating fins) are spaced along trunks to boost the signal. A CATV POWER SUPPLY provides the necessary power for all the amplifiers within a large radius. TAPS in the distribution lines allow the connection of DROPS, which provide service to each individual customer. CATV trunks and distribution lines are easily recognizable by their EXPANSION LOOPS. These loops are present because coaxial cables are rigid, and they expand and contract with changes in temperature at a different rate than the messenger wire. Without room for thermal movement, they can suffer undue stresses and deterioration and even pull themselves out of connections.

Both cable and telephone providers now often use fiber-optic cables in combination with copper wires or coaxial cables to distribute higher quality and more reliable signals. These cables utilize bundles of glass or plastic fibers to transmit signals as pulses of light. Fiber-optic signals can travel huge distances with very little loss because they are immune to electromagnetic interference. The outside of the cable sometimes includes an orange or yellow MARKER or overwrap, making it easy to distinguish from telephone or CATV.

Fiber-optic networks are usually designed to allow for future expansion by including more fibers than needed. However, a major challenge with these cables is that they are difficult to splice. Rather than a simple physical connection required for an electrical splice, fiber-optic cables need much more care to avoid scattering or reflection of the light signal. The individual fibers must be stripped, cleaned, cleaved, aligned, and precisely joined, often using heat to fuse them together. Instead of performing this careful operation on a ladder or in a bucket truck, many utilities prefer to add new connections or repair their fiber-optic cables in specialized *splicing trucks*.

That means the cables need enough slack to be lowered to the ground, and these SLACK LOOPS are usually stored along the main cable. Fiber-optic cables cannot have sharp bends or twists, which could break the fibers, so STORAGE BRACKETS (often simply called *snowshoes* for their distinctive appearance) allow changes in direction and storage of slack without causing damage to the cable.

KEEP AN EYE OUT

The electrical signals used in copper wire telephone systems are relatively small, so they usually cannot travel over long distances. That means almost all of us live and work within a few miles of a local telephone exchange. These days, most telephone switching happens on server racks in data centers, but many of the original exchange buildings still exist. These buildings, also called central offices, are owned by the service provider to house the equipment and switches that connect individual lines to the massive telephone network. They are usually generic and windowless, difficult to notice unless you're paying close attention. A few hints are the security cameras, air conditioners to keep all that equipment cool, and backup generators to power the system in the event of a blackout.

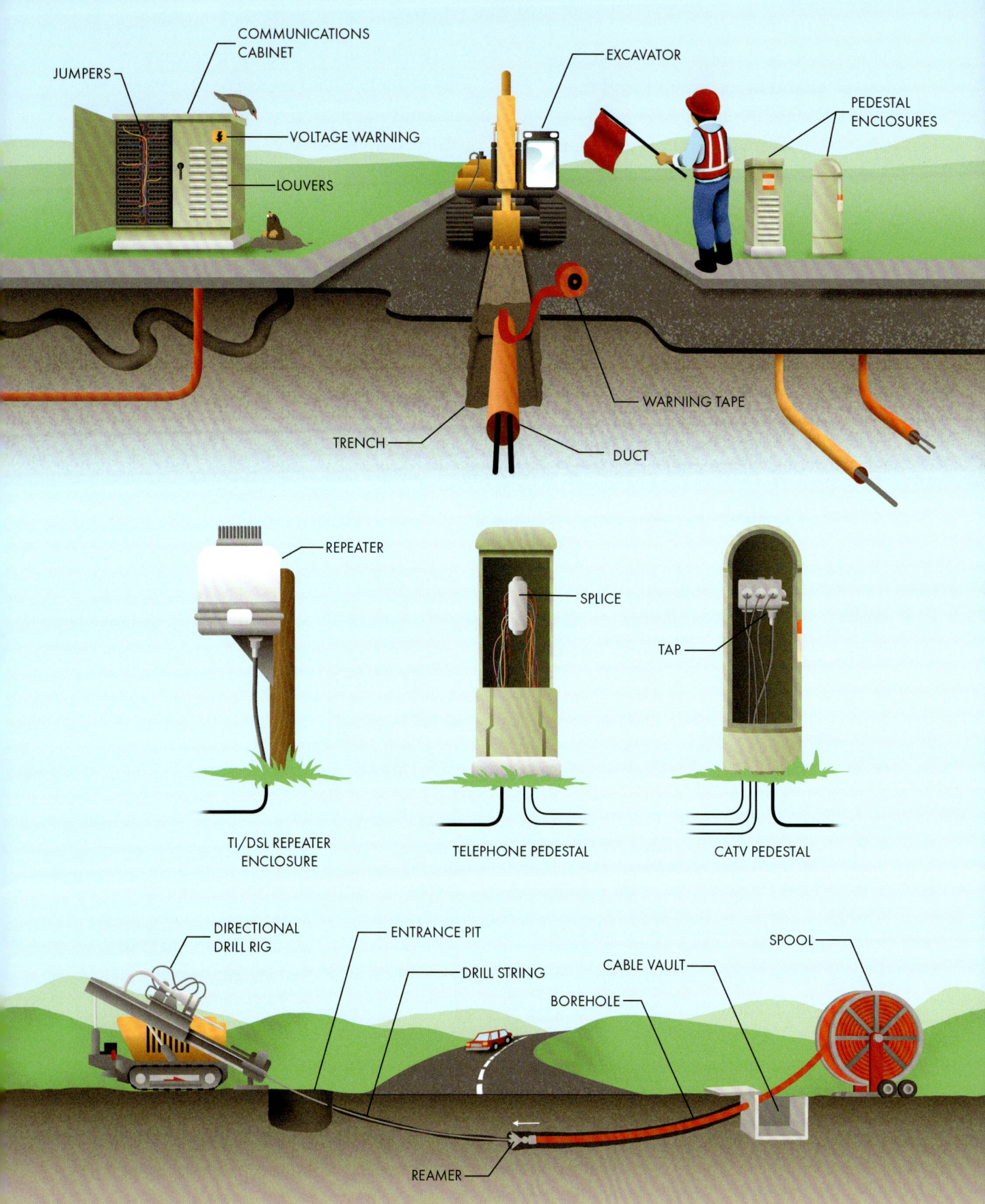

Underground Telecommunications

Routing communication lines underground rather than overhead along utility poles has some serious advantages. The lines don't require a supporting strand to bear their weight across poles. They are also less obtrusive and avoid adding visual clutter to the landscape. Finally, they are protected from a whole host of threats including birds, squirrels, wind, ice, sunlight, and errant vehicles crashing into poles. That means underground communication lines are usually more reliable, even though they require higher upfront costs for installation.

Underground utilities are usually placed inside protective DUCTS installed in one of two ways: *trenching* or *directional boring*. Trenching is accomplished using an EXCAVATOR to dig a linear hole (that is, a TRENCH) in the ground. The duct is placed within the hole, which is then backfilled with soil. WARNING TAPE is installed while backfilling to mark the presence of the cable to anyone who may be excavating around the line in the future. Some of these tapes even include wires or steel ribbons so that they can be detected from the ground surface, making it easier to locate the line in the future. The major disadvantage of trenching is the disruption caused to anything on the surface. The area must be closed off during construction, and sidewalks, roadways, and lawns have to be repaired after backfilling the trench. These repairs never seem to be as durable or attractive as the original.

Directional boring reduces the disturbance on the surface by installing ducts inside a BOREHOLE without a trench. This method is especially advantageous for getting lines across rivers, congested urban areas, and critical roadways where it wouldn't be feasible to use a trench. A DIRECTIONAL DRILL RIG on the surface first bores a pilot hole between the ENTRANCE PIT and the exit. Workers use sensing equipment on the DRILL STRING and at the surface to monitor the drill's path below-ground. To steer the drill, the leading edge of the string is asymmetric. It can be clocked to any position, and the string will naturally wander in the preferred direction during drilling. Once the pilot hole is complete, the drill string is retracted with a REAMER to enlarge the bore while pulling the duct from a SPOOL, creating a continuous path for cables to run.

Since they're hidden below the surface, you can't see underground telecommunication lines like you can with aerial installations. However, these cables have to surface eventually, so there are plenty of opportunities for spotting them. The simplest structure associated with underground utilities is a CABLE VAULT, a belowground enclosure that allows access to the ducts. Vaults are evident on the surface from their lids (which are often large rectangles and include details about what's inside).

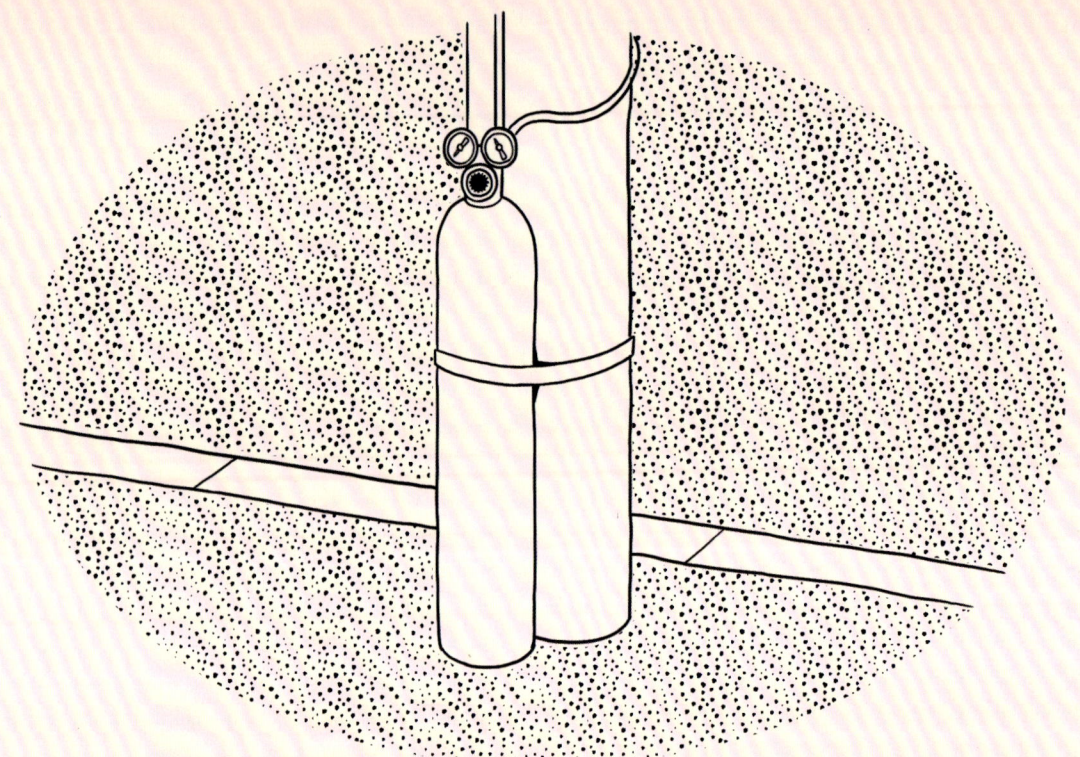

Buried cables have another major disadvantage besides cost: moisture. Rain, melting snow, and groundwater can find their way into the ducts that carry telecommunication lines below the ground. Not only does water cause corrosion if it gets inside a cable jacket, but it can also create short circuits and signal degradation. Moisture is mostly a problem for telephone lines (as opposed to coaxial or fiber-optic cables), because they have so many individual copper wires, and older cables were often insulated using paper. To counteract the intrusion of moisture, many telephone cables are pressurized with air inside the sheath using a compressor near the central telephone office. However, occasionally you'll see a nitrogen tank on a sidewalk or beside a street to pressurize underground lines. This pressurization helps the sheath resist the intrusion of water. Also, by monitoring the pressure, technicians can find and diagnose problems with the line before serious deterioration occurs. Any breaks or holes in the line will allow the air or nitrogen to leak and the pressure to drop over time. Most newer telephone cables are filled with a water-repellant gel, but plenty of these air-filled underground lines serve as a testament to the clever use of pressure for preventive maintenance.

Another structure associated with underground telecommunications is the COMMUNICATIONS CABINET. These cabinets are located aboveground and may house a wide variety of equipment for many different types of service providers, so you have to be a sleuth if you want to know exactly what's inside. The first clue is the labels. Sometimes you can find a company name or contact information on the cabinet giving a hint as to what kind of equipment it contains. Usually, cabinets serve as simple connection points, conveniently accessible locations for splicing a high-capacity trunk or feeder cable to smaller distribution lines that spread out toward customers. In such a case, the cabinet houses JUMPER equipment allowing technicians to make connections for CATV, telephone, or fiber-optic lines.

Some communications cabinets house *active* (in other words, powered) equipment. In these cases, a VOLTAGE WARNING likely will be somewhere on the outside, and the housing will have LOUVERS, since these devices often need ventilation to dissipate heat. Active equipment could include power supplies for the CATV network or *optical nodes* that convert fiber-optic signals into radio frequencies that can be distributed using coaxial cables.

Finally, these cabinets occasionally contain more sophisticated equipment that allows a telephone line to transmit information at much higher speed and fidelity than it could if it were connected directly to the nearest central office. These devices, called *remote concentrators*, digitize the signals from individual telephone customers and combine them into a fiber-optic signal directly to the central office, allowing telephone companies to service a larger number of customers and provide higher quality voice and high-speed data services.

Another sign of underground communication lines is the PEDESTAL. These ubiquitous housings are usually termination points, providing the connection between a larger distribution line and smaller cables that fan out toward customers for CATV, telephone, or other telecommunication services. They usually include an access panel or allow the housing to be removed so that technicians can make connections or troubleshoot problems. For CATV, they may include a TAP to provide multiple service drops. For telephone, they usually conceal only cable SPLICES and not much else.

One last piece of equipment associated with underground utilities is the REPEATER. *T1* and *DSL* are two common types of high-speed digital signals that can be transmitted along standard copper telephone lines. However, because of their high frequency compared with voice signals, these high-speed digital signals cannot travel far without becoming too attenuated or distorted. In rural areas with longer distances between telephone offices, these lines need repeaters to maintain the signal fidelity. The repeaters are normally housed in waterproof enclosures shaped like paint cans or crock pots. They'll show up at regular intervals along the line, usually every mile or two.

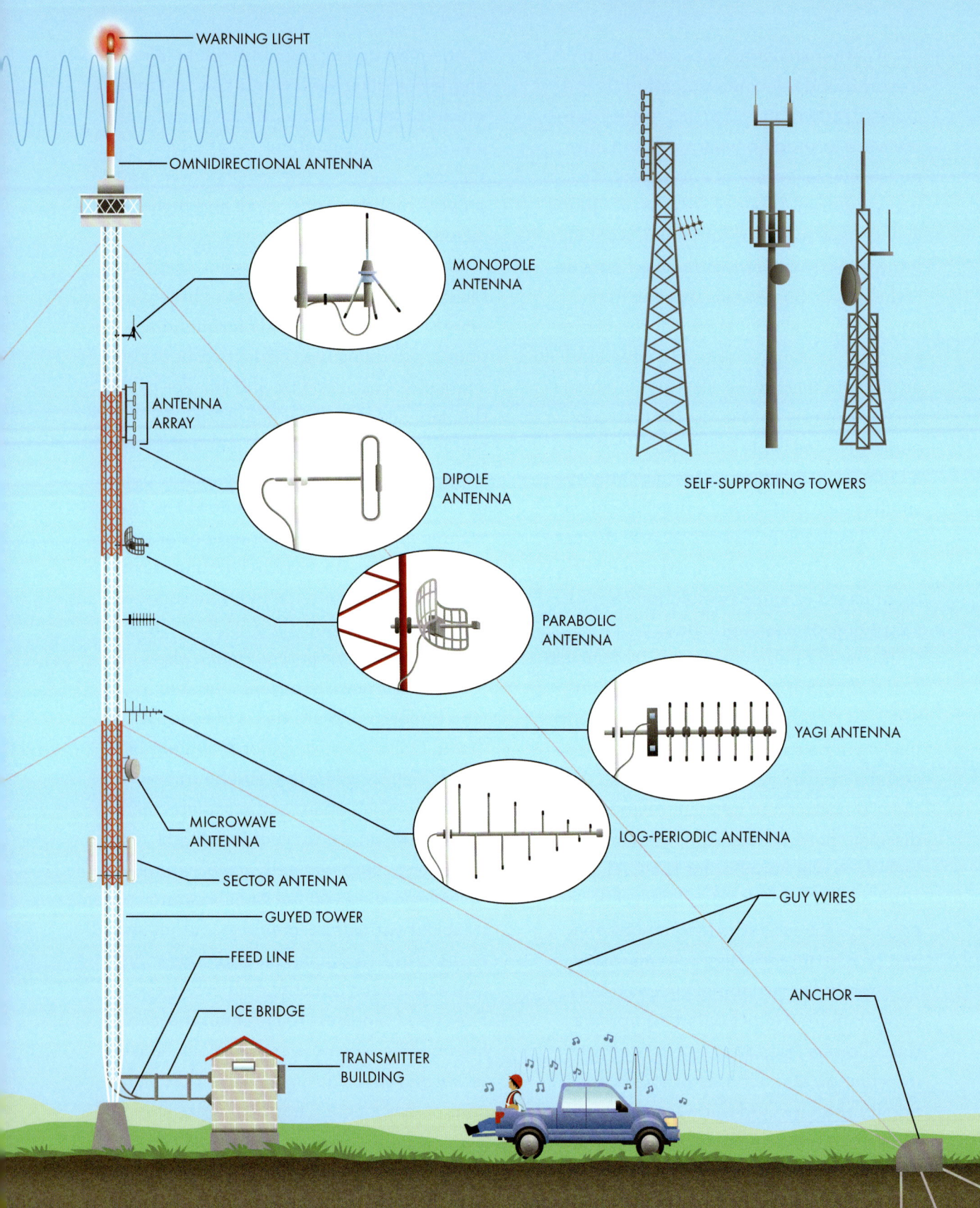

WARNING LIGHT

OMNIDIRECTIONAL ANTENNA

MONOPOLE ANTENNA

ANTENNA ARRAY

DIPOLE ANTENNA

PARABOLIC ANTENNA

YAGI ANTENNA

LOG-PERIODIC ANTENNA

MICROWAVE ANTENNA

SECTOR ANTENNA

GUYED TOWER

FEED LINE

ICE BRIDGE

TRANSMITTER BUILDING

GUY WIRES

ANCHOR

SELF-SUPPORTING TOWERS

Radio Antenna Towers

Radio communication uses invisible waves of electromagnetic radiation to carry information across space. This simple but remarkable technology enables a wide variety of wireless devices, from garage door openers to cell phones. If humans could sense the full spectrum of electromagnetic radiation, we would be completely overwhelmed by the volume and variety of information moving through the airwaves.

Many of the frequencies used for communication, including those broadcast by radio and television stations, require a line of sight; the path between the transmitter and receiver must be relatively unobstructed. Radio signals generally can't reach beyond the horizon, which is why many *antennas* are mounted at the tops of gigantic TOWERS (sometimes also called *masts*). The higher they are, the farther their signals can extend. Antenna towers are some of the tallest human-made structures in the world, with many topping out above 600 meters (about 2,000 feet). They are so tall that they often pose a danger to aircraft and often must be painted with alternating bands of orange and white color and feature WARNING LIGHTS at the top. These towers serve a critical role in modern society, enabling wide transmission of radio and television signals, communications for emergency first responders, and more.

Radio antenna towers can take many forms, but there are two main categories of structure (not including the spires atop tall buildings): SELF-SUPPORTING and GUYED. Self-supporting towers are designed to be freestanding and stable against the wind entirely on their own. They are usually made from steel or concrete with a broad base to provide stiffness against the forces of Mother Nature. Self-supporting towers don't take up much space, so they are ideal in urban areas where land comes at a premium. However, they are more expensive to construct than their counterpart because of the extra material required for stability against lateral wind loads.

Guyed towers usually consist of a slender lattice structure supported by multiple steel cables (the GUY WIRES). Guyed towers can be thin because they don't have to provide stiffness against the force of the wind. The guys provide this lateral support so that the tower has to support only its own weight. In fact, some guyed towers come to a narrow point at the ground so that any minor swaying will cause only a pivot instead of bending or flexing of the tower. The guy wires are usually arranged in an equilateral triangle to provide support no matter which direction the wind blows.

There are many ways to anchor the guy wires to the ground, depending on the type of soil or rock at the site and expected loads. The ANCHORS often consist of one or more deep drilled holes with a steel rod grouted inside to create a rigid connection to the earth. Because the guy wires extend so far from the tower base, guyed towers require much more space than self-supporting structures. They are mostly located in rural areas where land is less expensive.

Entertainment programming or other signals arrive at a tower site for broadcast from a radio transmitter. The transmitter is usually located away from the tower inside an environmentally controlled TRANSMITTER BUILDING. For AM radio stations, the tower itself is the antenna and there may be a tuning hut at the tower base that houses equipment needed to efficiently transfer power from the transmitter to the tower. For FM and TV stations, the FEED LINE (also called a transmission line) carries the signal from the transmitter up to the antenna, which is attached to the tower structure. In colder areas, the horizontal run of feed line from the transmitter building to the tower is protected from falling ice by an ICE BRIDGE. The *antenna* is the device that radiates the signal as electromagnetic waves. Since towers are fairly expensive and obtrusive, they are often shared among many stations or other users (called *colocation*). Tower owners lease space inside the transmitter building and on the tower structure to radio and television stations, police and fire departments, government agencies, and various private companies for their own wireless communication systems.

Like the towers to which they're attached, these antennas can take a wide variety of interesting shapes, depending on the frequency, direction, and power of the signal. OMNIDIRECTIONAL ANTENNAS transmit radio waves in all directions equally and often have a cylindrical shape. They include MONOPOLE ANTENNAS,

straight conductive elements that require a ground plane (sometimes the ground itself and sometimes consisting of radial, horizontal conductors). DIPOLE ANTENNAS are another kind of omnidirectional antenna that consist of two identical radiating elements, one above the other.

Directional antennas focus radio waves in a specific direction. PARABOLIC ANTENNAS have a solid or gridded wire dish to reflect and focus radio waves. YAGI ANTENNAS use a single energized dipole and several non-energized elements to focus the waves in the desired direction. Similar in appearance, LOG-PERIODIC ANTENNAS use a series of dipoles, each of a slightly different length to send or receive a wide range of radio frequencies. Simple antenna elements, such as dipoles, can be combined into ARRAYS that work together to direct waves into a beam or a specific pattern. (A few other types of antennas, including those used for cellular phone service, are discussed in another section.)

Like all infrastructure, antenna towers require occasional maintenance. Technicians with specialized training for heights and electrical hazards perform inspections and upkeep on these structures. Very tall towers may be equipped with elevators to provide access for painting, repairs, and equipment changes. Shorter towers require the technician to climb to the top.

Although the frequencies used for wireless communication are *non-ionizing* (meaning the waves can't break apart atoms), that doesn't mean they aren't dangerous. Electromagnetic radiation can generate heat in water-containing objects, including people. (Microwave ovens take advantage of this effect to warm our food.) That's why public access is restricted near antennas that transmit at high power.

Workers who maintain these towers have to make sure to keep their distance from energized antennas or power them down before working nearby to avoid unsafe exposure.

KEEP AN EYE OUT

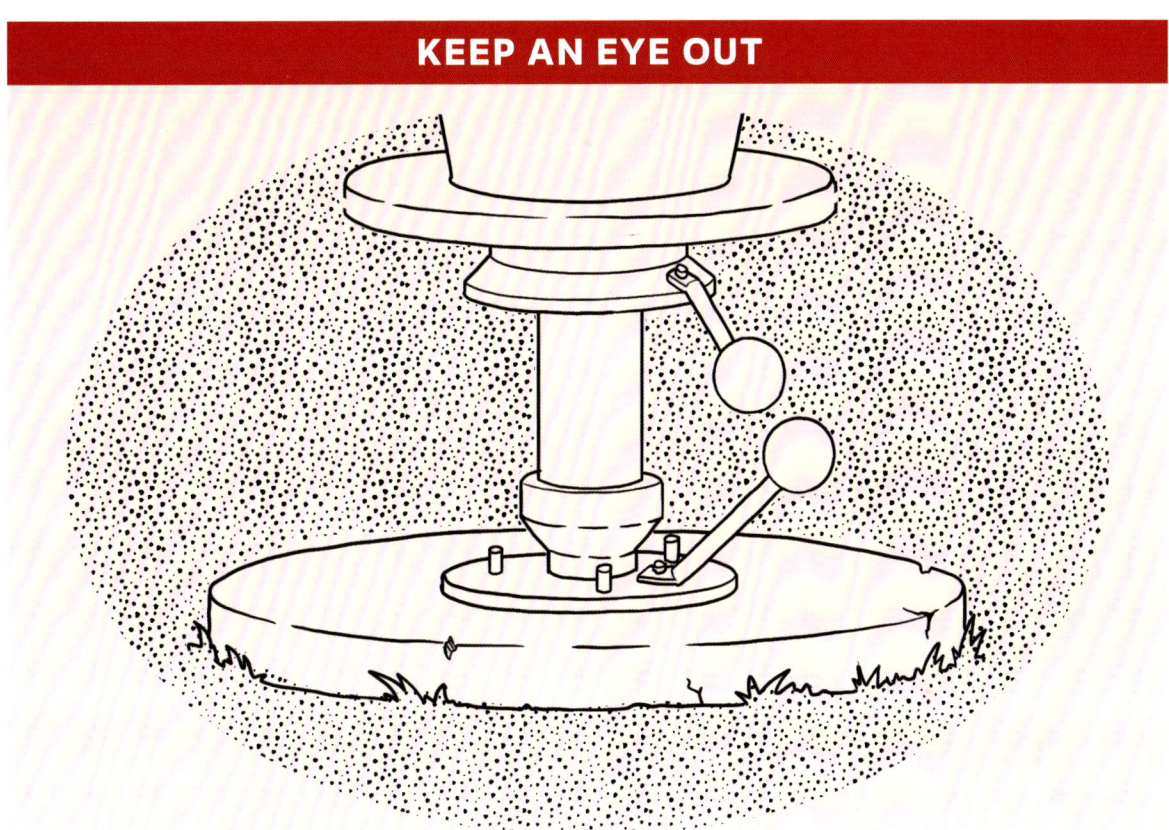

AM radio signals use very low frequencies, so they require very large antennas. In most cases, an AM station will broadcast using the metal tower itself as the antenna. Because the entire tower is energized, it has to be insulated from the ground. If you look closely, these towers usually sit entirely atop a small ceramic insulator. The need for complete isolation from the ground creates a number of interesting challenges, one of which is how to protect the tower and the attached equipment against damage due to lightning. Many AM towers use *spark gaps* to keep the tower insulated while allowing voltage surges to be diverted safely into the ground. During normal operation, no current conducts across the gap. However, if lightning strikes the tower, the air will ionize between the contacts, creating an arc, and providing a conductive path to ground for the voltage spike.

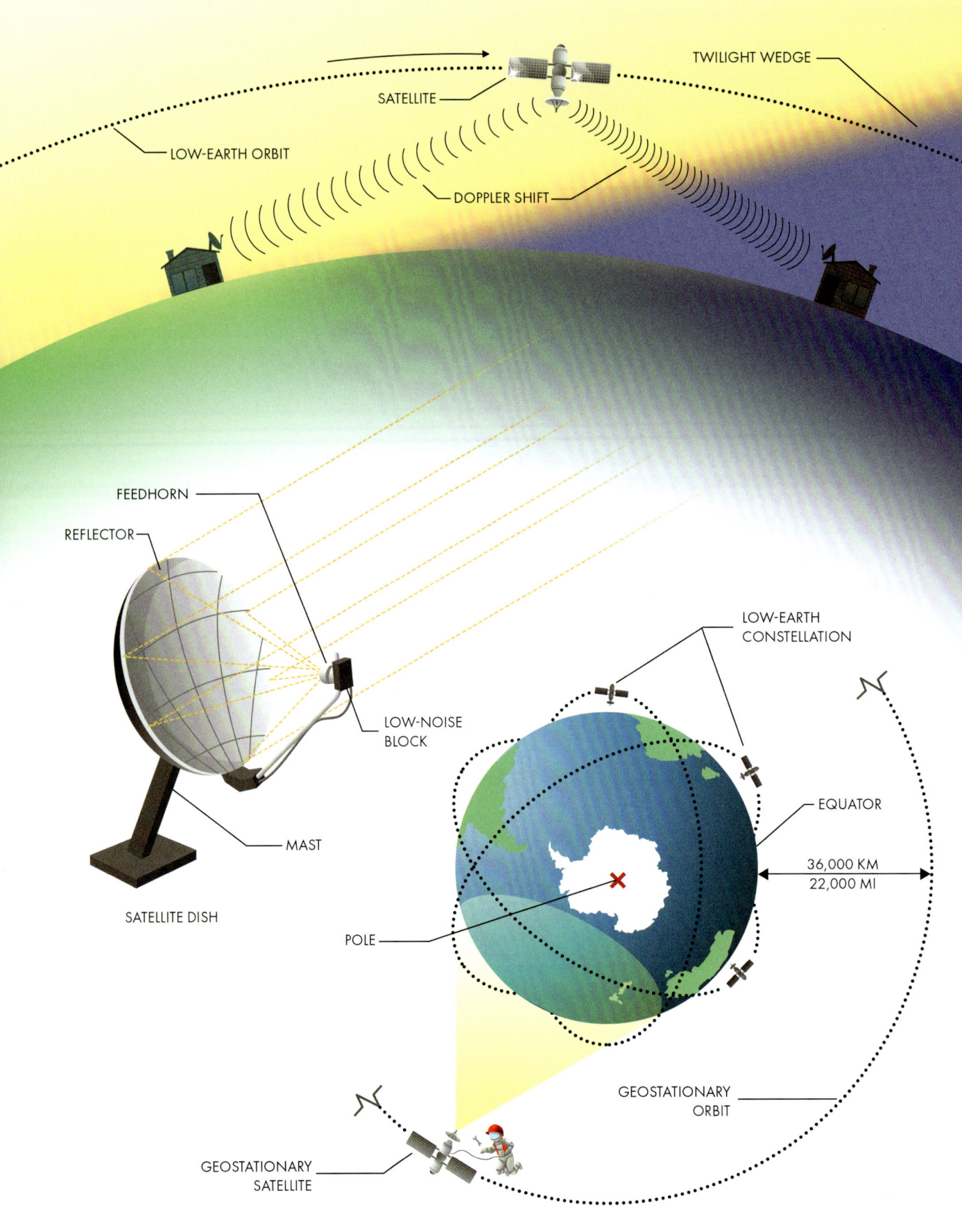

SATELLITE

LOW-EARTH ORBIT

TWILIGHT WEDGE

DOPPLER SHIFT

FEEDHORN

REFLECTOR

LOW-NOISE
BLOCK

MAST

SATELLITE DISH

LOW-EARTH
CONSTELLATION

EQUATOR

36,000 KM
22,000 MI

POLE

GEOSTATIONARY
ORBIT

GEOSTATIONARY
SATELLITE

Satellite Communications

There is a practical limit to the height of an antenna mast. Eventually, the financial, engineering, and safety challenges make it infeasible to build taller. Luckily, there is another way to get an antenna high in the sky. SATELLITES are devices placed into orbit around the Earth using rockets. They are the pinnacle of wireless communications, at least in terms of range. Many satellites can transmit and receive radio signals from one-third of the globe simultaneously, considerably farther than even the tallest towers. These days we use satellites for a wide variety of communications, including radio, television, internet, telephone, navigation, weather, environmental monitoring, and so much more. Satellites used for communications are essentially relays, receiving signals from one location on the ground and amplifying and redirecting them back to somewhere else on Earth. This relay creates a communications channel that does not require a direct connection through wires and that isn't so limited in range by the curvature of the Earth like ground-based antennas.

Communication satellites can be placed into a wide variety of orbits around the Earth. The speed at which a satellite orbits is directly related to its *altitude*. The higher the orbit, the longer it takes to go around. Satellites in LOW-EARTH ORBIT circle the globe many times per day, so they are overhead a specific location for only short periods of time. A group of satellites in overlapping orbits, called a

CONSTELLATION, is required to maintain continuous service. Each satellite is strategically placed so that any location on the ground has at least one satellite within line of sight at all times. Low-Earth orbiting satellites require less power to transmit and receive, and communications experience less delay, since they are closer to the Earth. They also do not require a large antenna to receive their signals. In fact, you probably carry an antenna in your pocket that connects to low-Earth satellites: the *GPS* antenna in your cell phone. However, low-Earth satellites do have to account for DOPPLER SHIFT. Because the satellites move so quickly compared to an observer on Earth, the radio waves compress while moving toward an antenna and stretch out as they pass by overhead, complicating the job of receiving and decoding the signals.

At an altitude of approximately 36,000 kilometers (22,000 miles), the *orbital period* of a satellite is 24 hours, exactly the length of a day. A satellite at this altitude around earth's EQUATOR is in a GEOSTATIONARY ORBIT because it remains in a fixed position in the sky as the Earth rotates. Although it takes considerable effort to launch into an orbit so high above the Earth, GEOSTATIONARY SATELLITES have some significant advantages. Because they don't move relative to the ground, antennas can be mounted in a fixed position, simplifying their design. Geostationary satellites also have a much larger range since they have a line of sight that covers about 40 percent of the globe. Only the Earth's POLES are difficult to reach from this orbit.

One limitation of geostationary satellites is that they are constrained to a single ring (called the *Clarke Belt*) above the Earth's equator. To avoid satellites interfering with each other's signals, the international telecommunications community agreed to designate individual locations (called slots) around this ring like parcels of real estate. The geostationary ring is so congested that it has a waiting list. Once a satellite has reached the end of its useful life, it must move out of its slot so that a replacement or a new satellite on the waiting list can take its place.

The other disadvantage with geostationary satellites is their large distance from the Earth. Sending and receiving radio signals across this great expanse is a major challenge. The antennas used to overcome this distance are instantly recognizable. A SATELLITE DISH uses a curved REFLECTOR to gather the faint radio signals and focus them into the FEEDHORN. This metal cone transitions the waves into the LOW-NOISE BLOCK, the heart of the satellite antenna that includes electronic circuitry to perform two primary functions. First, it amplifies the weak radio signal to a more usable level. Second, it takes the high-frequency signal used for long-distance wireless transmission and *downconverts* it to a lower frequency that can travel efficiently through a cable.

Antennas that transmit signals to geostationary satellites are usually much larger but otherwise work in the same way, with equipment for amplifying and converting the frequency and a reflector to direct the waves to the correct location in the sky. The MAST supporting the dish can be attached to a permanent mount or motorized tracking mount, depending on whether it communicates with only one or multiple geostationary satellites.

Some satellites are large enough and reflective enough to be seen from the ground at night. In fact, with so many orbiting the Earth these days, spotting satellites is a popular hobby. Many websites keep track of satellite orbits and offer predictions about when and where they might be seen and how bright they will appear in the sky. That brightness comes from glints of sunlight reflected off solar panels or shiny surfaces down to Earth, which is why satellites are most visible for a few hours right after nightfall or right before dawn. During those times, the sky is dark from Earth's shadow (sometimes called the TWILIGHT WEDGE), but the sun is close enough to the horizon to illuminate objects high above the ground. The most famous satellite orbiting Earth, the International Space Station, is also the largest and most visible. In most parts of the world, you're likely to be able to see this feat of modern engineering zoom across the night sky at least a few times per month. It is a spectacular sight.

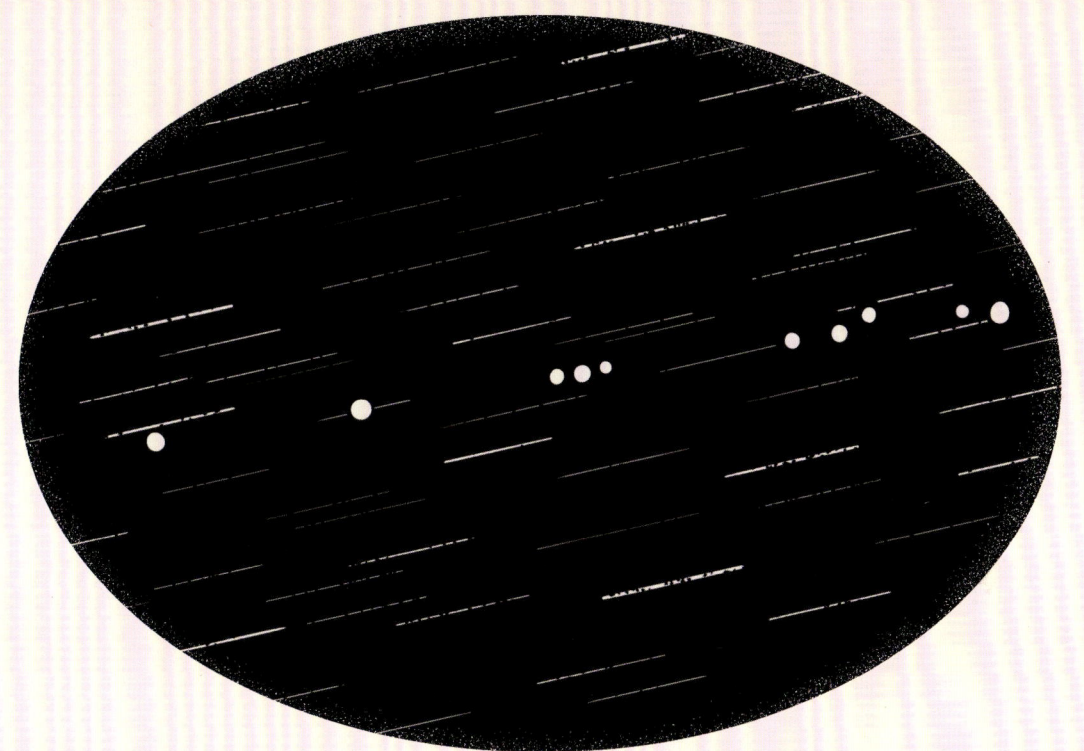

Because they orbit much farther from Earth, geostationary satellites are illuminated by the sun all night long. However, that distance also means they appear much dimmer in the night sky. Usually, these satellites can only be seen with a telescope, but there's another clever way to observe them: long-exposure photography. Point a camera on a tripod at the celestial equator and open the shutter for two to four minutes. In the resulting photograph, you'll see the long trails of stars caused by rotation of the Earth. But, if you look closely, you should see a row of pinpoint lights. These are the geostationary satellites orbiting at the exact speed of the Earth's rotation, so they always appear in the same part of the sky.

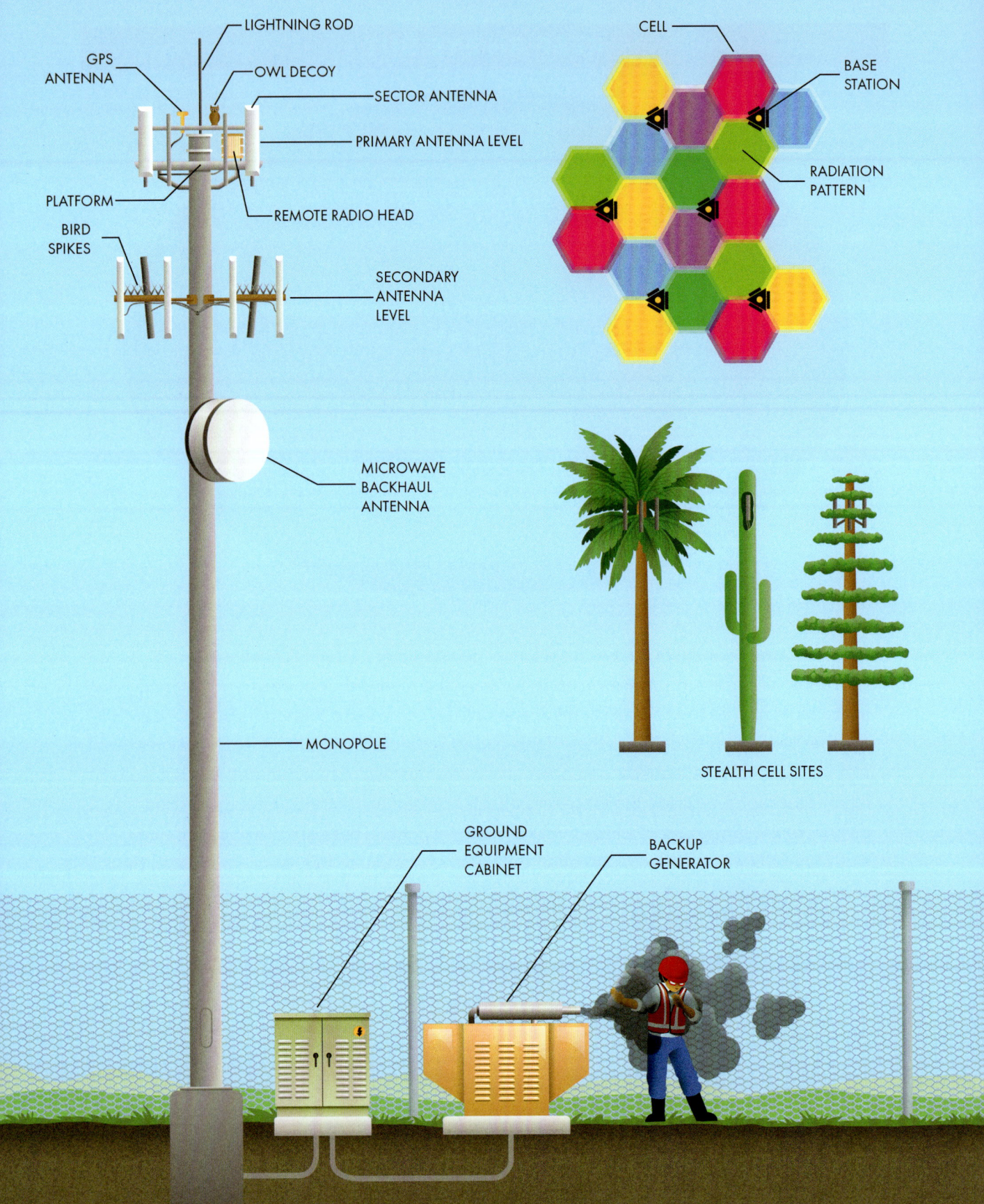

LIGHTNING ROD

GPS ANTENNA

OWL DECOY

SECTOR ANTENNA

PRIMARY ANTENNA LEVEL

PLATFORM

REMOTE RADIO HEAD

BIRD SPIKES

SECONDARY ANTENNA LEVEL

MICROWAVE BACKHAUL ANTENNA

MONOPOLE

GROUND EQUIPMENT CABINET

BACKUP GENERATOR

CELL

BASE STATION

RADIATION PATTERN

STEALTH CELL SITES

Cellular Communications

Most wireless communications involve either one-way broadcast of a signal (for example, AM and FM radio) or two-way transmissions between a limited group (such as a police dispatch network). The availability of different frequencies on the electromagnetic spectrum used for separate "channels" of communication is limited. On top of that, there's a lot of competition for those limited bands among the wide variety of radio signal users, including public safety organizations like police and fire departments, military, aircraft traffic control, television and radio stations, and many more. Enabling wireless telephony and internet connectivity to the greater public is a major engineering challenge. Within only a narrow range of frequencies, wireless carriers have innovated ways to connect anyone with a mobile device to both the telephone network and the internet. The fundamental innovation making this possible is the subdivision of large service areas into smaller CELLS—hence the name, cellular communications.

Although it would seem more economical to mount communication antennas atop tall towers to reach the largest possible area, that would allow only a few connections at a time (one per channel within the available band of radio frequencies). Instead, carriers install many smaller antennas spread out across the landscape to service manageable groups of customers. This strategy allows billions of individual wireless transmissions per day on only a few hundred channels, since nonadjacent cells can reuse the same channels (shown as different colors in the illustration). Each cellular service carrier builds its own grid of cells providing coverage to all but the least traveled areas. Although idealized as a regular hexagonal grid, each cell's size and shape is determined by the topography, availability of antenna mounting locations, and especially, demand for service. Densely populated areas have smaller cells, while the cells in rural areas can be much larger.

The creation of all these cells has left a mark across the landscape in the form of BASE STATIONS. A base station (also known as a *cell site*) has all the infrastructure needed to provide service to one or more wireless cells, usually including a tower, antennas, amplifiers, signal processing equipment, a backhaul connection to the network, and sometimes batteries or a BACKUP GENERATOR for power outages.

The ubiquitous towers used to mount the antennas are a familiar sight. In urban settings, they are usually MONOPOLES or lattice structures. Often, the signal processing is done in a REMOTE RADIO HEAD located adjacent to the antennas, and other times the radio equipment is located in an EQUIPMENT CABINET on the ground. LIGHTNING RODS protect the sensitive equipment from strikes. The antennas also need deterrents to prevent damage by wildlife. If you look closely, you'll see a wide variety of creative approaches to this challenge. The most common are predatory DECOYS (usually owls) that scare birds away or plastic BIRD SPIKES that make climbing or roosting on antennas difficult. Another item you may notice on a tower is a GPS ANTENNA. This antenna is usually shaped like an egg and collects an accurate clock signal from overhead satellites required for synchronization of the signal processing equipment.

Base stations aren't always standalone towers, though. Keep your eyes open in urban areas, and you'll notice antennas on just about any tall structure, including buildings, water towers, utility poles, and even billboards. In fact, a highly developed economy exists around leasing space for cell site installations, complete with agents, investment firms, and all the other players in the traditional real estate market. Often, carriers will share a tower or a building to save cost and reduce the visual impact on the landscape of this conspicuous type of infrastructure. You'll often see two or more ANTENNA LEVELS on the same tower. Another way to curb a cell tower's obviousness is to disguise it as something more natural like a tree or cactus. Some of these so-called STEALTH CELL SITES are more stealthy than others.

These days, you're almost always within view of a set of the rectangular SECTOR ANTENNAS used to send and receive the signals used by mobile devices. These antennas are highly directional to maintain clear boundaries between cells, usually targeting a 120-degree swath of territory. The triangular PLATFORMS atop some towers allow for antennas to service three cells from one station, and each antenna is carefully aimed to avoid interference with neighboring cells. You may notice some antennas are tilted downward to reduce the spread of signal beyond the cell boundaries. The RADIATION PATTERN of each antenna sector is roughly circular. When you account for the necessary overlap to allow for the digital handoff if a device moves from one cell to another, you get a roughly hexagonal grid.

The connection of each base station to the core network is called the *backhaul*. In most cases, a cellular base station's backhaul is accomplished using a fiber-optic cable to the nearest switching center. In instances where fiber installation isn't feasible, carriers can use a wireless backhaul. The circular protrusions you occasionally see from cell towers shaped like bass drums are actually high-capacity MICROWAVE ANTENNAS. Below the protective covering is a parabolic dish similar to those used for sending and receiving signals from satellites. These antennas are directional. If you could look down the center of one, you'd see its pair mounted on a tower in the distance facing directly back at you.

Cellular infrastructure is probably the most rapidly evolving of all the topics covered in this book. What started as a means to provide mobile telephone service is now the primary access to the internet for many. Voice conversations have become a secondary feature of a mobile phone to the point that many prefer the term "device" over "phone." As more and more gadgets gain internet connectivity (often described as the *Internet of Things*), demand for high-speed wireless service is only expected to increase. Wireless carriers will have to continue to innovate, and that means the cellular infrastructure of today may not look much like that of tomorrow.

KEEP AN EYE OUT

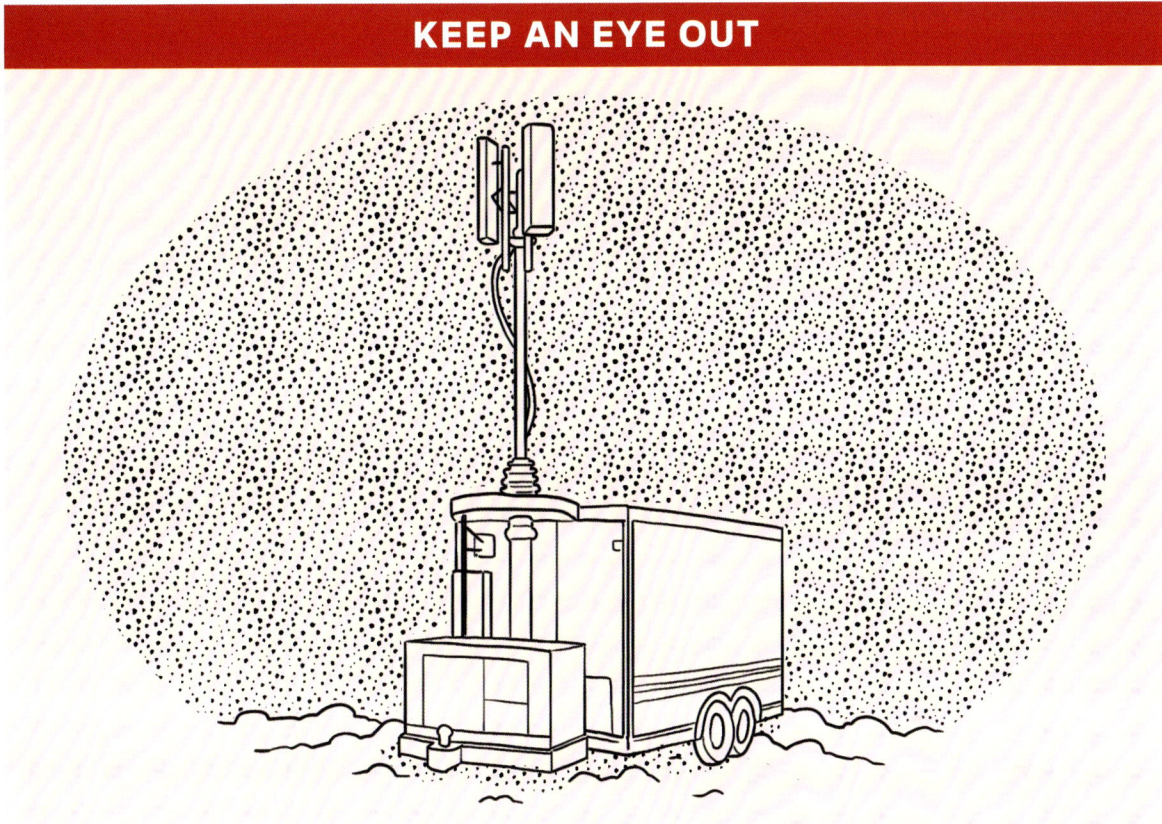

During major events like sports games and concerts, the demand on a cellular network can far exceed its capacity. Also, disasters and emergencies can disrupt existing communication networks when they're needed most. Mobile cell sites allow for on-demand expansion of cellular networks to add capacity or temporarily expand service into new areas. Affectionately known as *COWs* (for "cell site on wheels"), these truck- or trailer-mounted towers can be rented for deployment at a moment's notice. Look for telescoping towers attached to a trailer or truck at the next major event you attend, and be thankful for cell service when you want to access your mobile ticket or send a video of the event.

3

ROADWAYS

Introduction

Of all the elements of our constructed environment, roads may be the least noticed, yet they are almost as fundamental as the air we breathe. You almost certainly arrived where you are now via a road, and one will likely take you wherever you go next. The first roadways in history were formed as people or animals followed the same trail long enough to erode a path between two points. They have always existed in some form, but they haven't always been safe, comfortable, or able to accommodate the enormous number and weight of vehicles that use our present system of roadways every day. Over the years, the demands on streets and highways have only increased as more people and goods are on the move. Their designs have evolved alongside this demand. It may not always seem so, but roads now carry more and heavier vehicles than they ever have in history. With their omnipresence, their value to society is easy to forget. But the engineers, contractors, and public works crews who study, design, construct, and maintain our roadways know how important they are in moving goods and transporting people. Whether or not you love how much roads dominate the landscape, you have to marvel at the fact that, in most parts of the modern world, anyone can take a bus, car, bike, truck, motorcycle, or scooter to go almost anywhere else with relative ease and comfort.

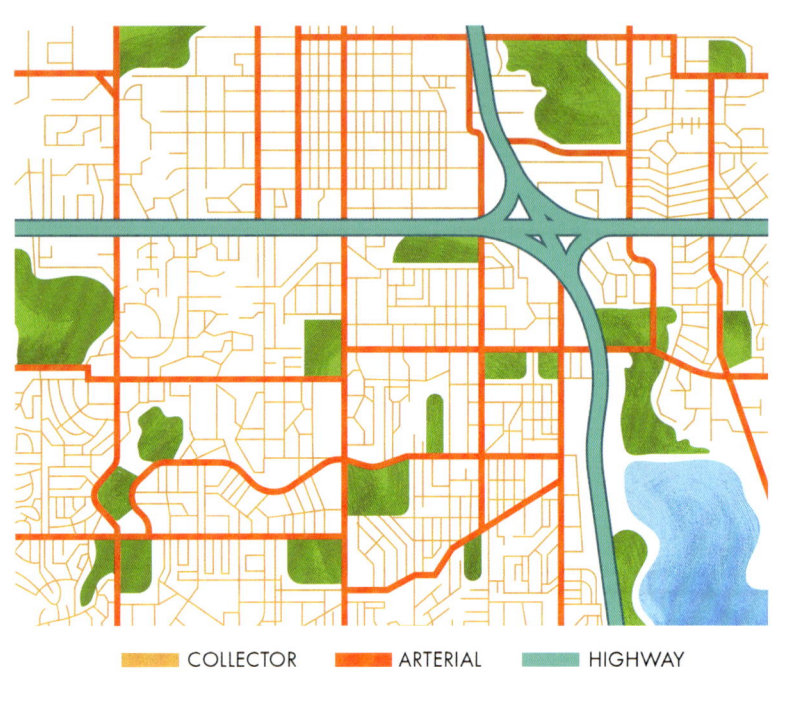

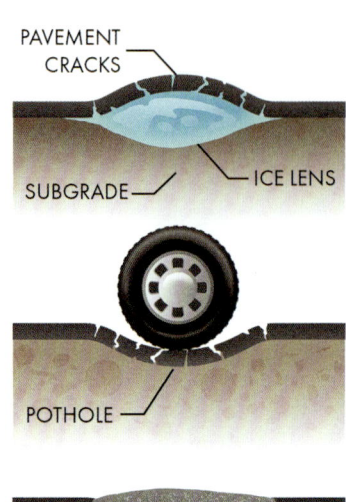

PAVEMENT CRACKS

SUBGRADE

ICE LENS

POTHOLE

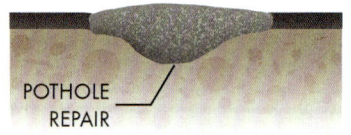

POTHOLE REPAIR

COLLECTOR   ARTERIAL   HIGHWAY

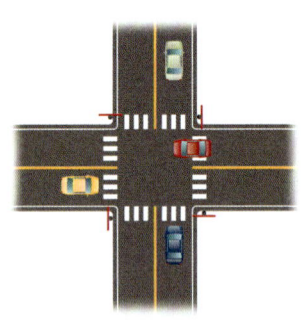

SIGN-CONTROLLED INTERSECTION

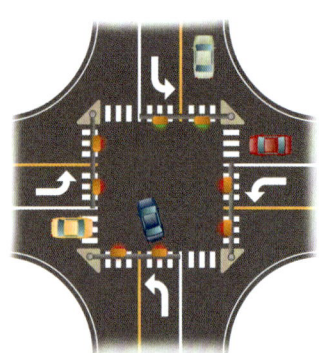

SIGNAL-CONTROLLED INTERSECTION

ROUNDABOUT

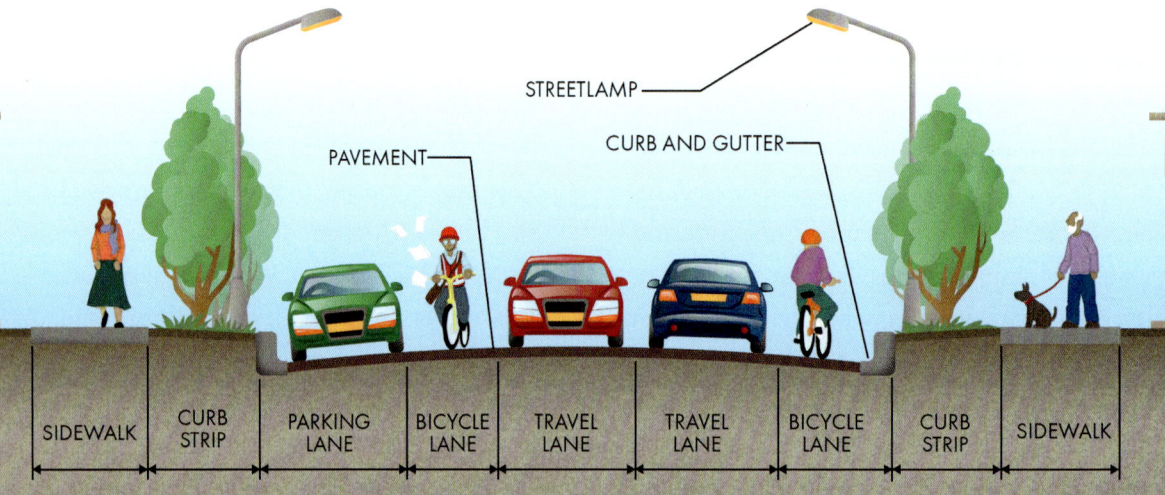

STREETLAMP

PAVEMENT

CURB AND GUTTER

| SIDEWALK | CURB STRIP | PARKING LANE | BICYCLE LANE | TRAVEL LANE | TRAVEL LANE | BICYCLE LANE | CURB STRIP | SIDEWALK |

Urban Arterial and Collector Roads

Nothing has had a more significant impact on the planning and design of cities in the past 100 years than the automobile. With their explosion in popularity in the early 20th century, motor vehicles became the standard mode of urban transportation. And with that, cities needed roads to accommodate the growing volume of traffic. There are a lot of good analogies between cities and human anatomy, and roadways are no exception. In fact, roads are often referred to by their cardiovascular equivalents. HIGHWAYS are like the aorta with a high capacity and single major destination. Small COLLECTOR roads are like capillaries with not much capacity but a connection to every individual house and business. In between are the aptly named ARTERIAL roadways, the medium-capacity connections between urban centers. All together, they form an urban transportation network allowing vehicles to travel (somewhat) efficiently between any two places on the map.

Although it might not always seem to be the case, streets in cities create routes for more than just automobiles. Collectors and arterial roadways truly make up the city's circulatory system, providing a path for cars, trucks, buses, bicycles, pedestrians, utility lines, and even stormwater runoff. Although every street is different, most urban roadways share many of the same features. This section provides an overview of the most common elements you might see in your city.

One way to characterize roadways is how they cross, also known as *intersections*. Collectors and arterial roads commonly overlap *at grade*, in other words, on the ground at the same level. That means only a few traffic streams can pass through simultaneously, resulting in an interrupted flow of traffic. These intersections are also where a vast majority of crashes occur. For those reasons, traffic engineers put much thought and analysis into the design of intersections and how to make them as safe and efficient as possible. This challenge almost always requires a compromise of numerous conflicting considerations, including space, cost, types and volumes of traffic, and human factors like habits, expectations, and reaction times. The simplest intersections are SIGN-CONTROLLED, using stop or yield signs to manage the flow of traffic. They are cost-effective and don't require any extra space, but they can't handle high volumes because they create an interruption for every vehicle passing through. SIGNAL-CONTROLLED INTERSECTIONS use electric lights to indicate which traffic can proceed. (Traffic signals are covered in detail later in the book.) ROUNDABOUTS are circular intersections that keep traffic flowing around a central island. Although they sometimes take up more space than other types of junctions, they have some distinct advantages. Roundabouts handle traffic efficiently by avoiding the start and stop of interrupted flow, and they create fewer dangerous collisions because of the slower speeds and single direction of traffic. Of course, within these three basic categories exist an endless array of intersection configurations. If you drive long enough, you will see the wide variety of intersection types and layouts used by engineers to keep roadway traffic flowing safely and efficiently.

Roadways consist of TRAVEL LANES for vehicles and occasionally space for BICYCLE and PARKING LANES as well. The roadway's surface is usually crowned in the center with a slope toward the outer edges to shed rainwater off the driving surface. At the outer edges, CURBS separate the PAVEMENT from the developed area, and GUTTERS provide a channel for rainwater to travel. Many cities and towns include a narrow strip between the road and SIDE-WALK to provide a safety buffer between fast-moving vehicles and vulnerable pedestrians. This area has many regional names, including CURB STRIP, *verge*, and *berm*. It also provides a location for utility poles, signs, and STREETLAMPS.

Unfortunately, pavement is not invincible. One of the most common frustrations of city driving is the POTHOLE. They're annoying, yes, but they're more than that. Potholes cause billions of dollars of damage to tires, shocks, and wheels of vehicles. Even worse, they're dangerous. Cars swerve to miss them, sometimes at high speeds, and if a bike, motorcycle, or scooter hits one, it can be bad news for the rider. The formation of a pothole happens in steps, the first of which is deterioration of the surface pavement. They might

seem innocuous, but CRACKS are critical flaws in a pavement system because they let in water. Soil below the pavement can become waterlogged from precipitation, softening and weakening the SUBGRADE. Water below the roadway can also freeze and grow into a formation called an ICE LENS. Water expands when it freezes, and it does so with tremendous force, separating the subgrade and pavement. When those lenses thaw out, the ice that was supporting the pavement recedes, creating voids. Every time a tire hits this soft area, it pushes some of the water and underlying soil back out of the pavement. It's a slow process at first, but every little bit of subgrade eroded from beneath the pavement means less support, and less support means more volume below the pavement for water to be pumped in and out by traffic. Eventually, the pavement loses enough support that it fails, breaking off and creating a pothole.

Because potholes are so destructive and inconvenient, roadway owners spend a lot of time and money both trying to prevent them from forming and fixing them when they appear. Prevention mainly involves sealing cracks against water intrusion. Repairs can be wide-ranging depending on the materials, cost, and climate conditions. But they all mostly do the same thing: replace the soil and pavement that was lost and (hopefully) seal the area off from further water intrusion. If the POTHOLE REPAIR does not create a good connection with the rest of the roadway, a pothole can recur in the same location.

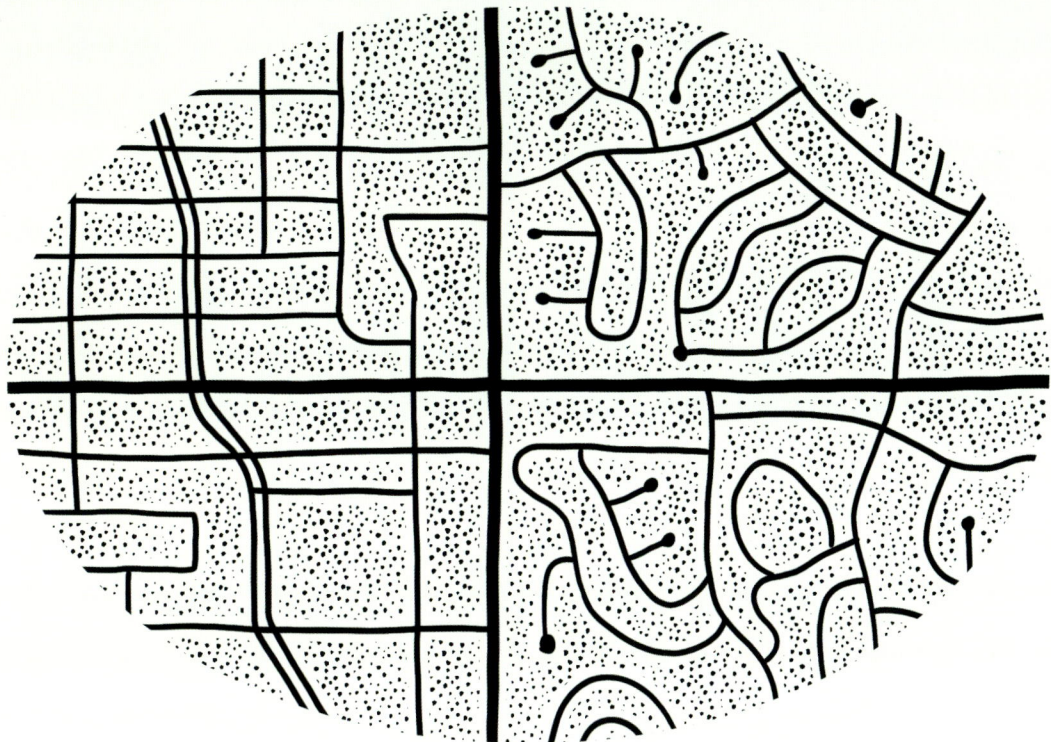

The layout of urban streets varies across the world and even across individual cities. Many cities are arranged in a rational grid pattern. The gridiron is as old as human history, and many of the earliest planned cities organized their streets at regular intervals and at right angles to each other. This pattern makes it easy to find your way and gives you many choices in selecting a route. However, it has some disadvantages. For one, it creates numerous intersections where crashes are most likely to occur. Also, grids make every street a *through* street, which often means more noise and drivers who may be less careful about their surroundings.

Many newer neighborhoods were designed to be disconnected from the main transportation networks to discourage through traffic. The streets are arranged in curvy loops, with T intersections, and *cul-de-sacs* to slow down traffic and reduce the number of crashes. Connections to main roads happen in only a few locations, which means the cars on the streets consist mainly of neighborhood residents who are more likely to drive carefully. This style of street layout is not without its disadvantages, though. The disconnected, circuitous routes sometimes make it difficult to use any transportation mode other than a motor vehicle. In many places around the world, modern neighborhood planning includes a focus on better connectivity for pedestrians, cyclists, and mass transit users.

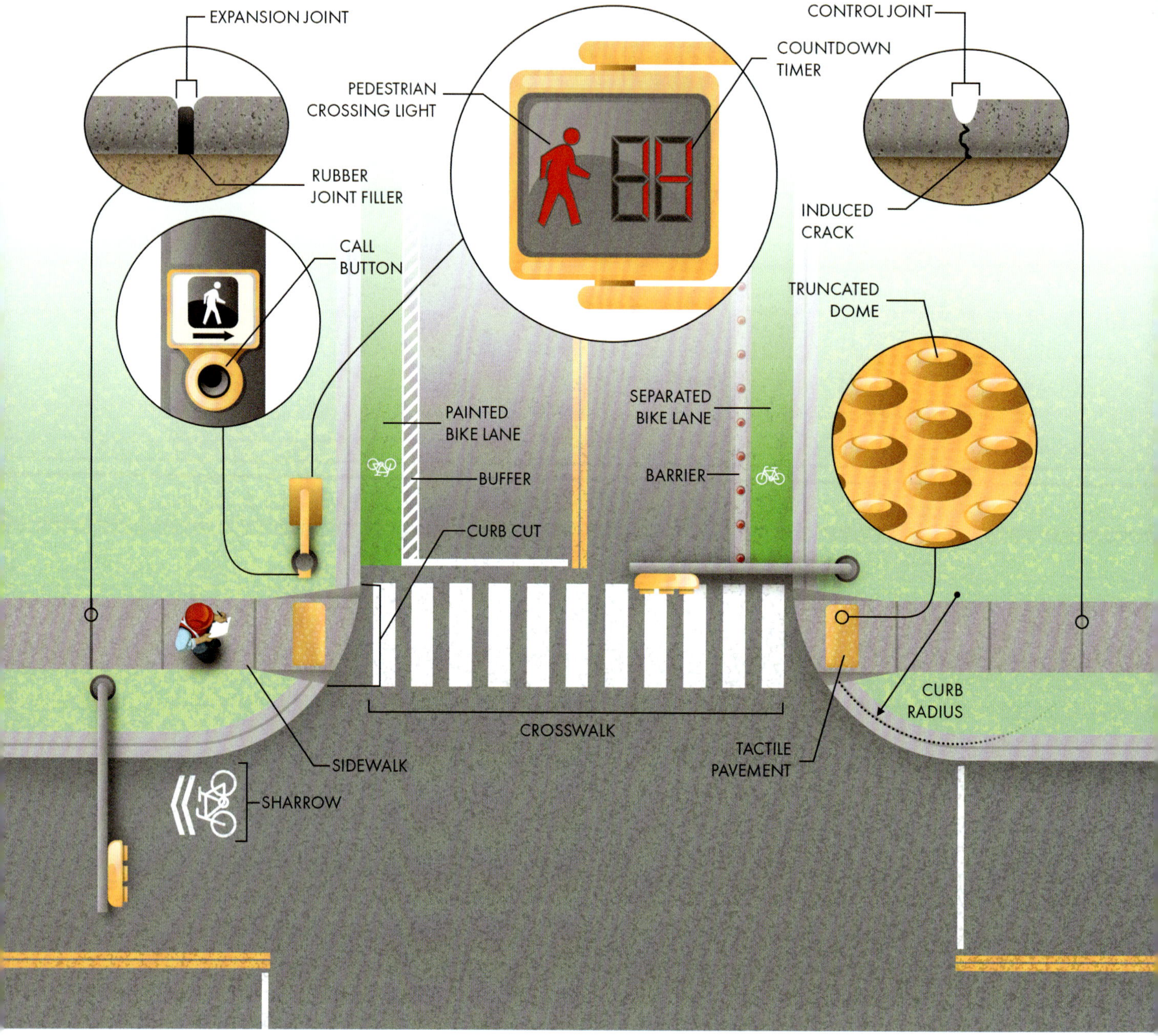

EXPANSION JOINT

RUBBER JOINT FILLER

CALL BUTTON

PEDESTRIAN CROSSING LIGHT

CONTROL JOINT

COUNTDOWN TIMER

INDUCED CRACK

TRUNCATED DOME

PAINTED BIKE LANE

BUFFER

SEPARATED BIKE LANE

BARRIER

CURB CUT

CROSSWALK

SIDEWALK

TACTILE PAVEMENT

CURB RADIUS

SHARROW

NECKDOWN

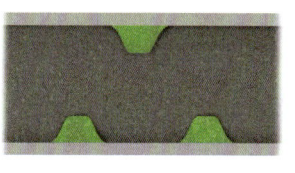

CHICANE

TREES

SPEED HUMP

TRAFFIC CALMING MEASURES

Pedestrian and Bicycle Infrastructure

Much of our present system of roadways was designed according to a single measure of performance: the safe and efficient movement of motor vehicle traffic. There was a time when cars were not so central to our urban lives. However, for the past 100 years or so, they seem to have been the principal consideration in every part of the planning and design of cities. Unfortunately, this car-centric approach deprives all the other users of urban roadways, including pedestrians and cyclists. In many places, if you try getting around town in anything other than a personal automobile, you'll face a string of inconveniences and dangers along the way. Luckily, cities are realizing the importance of walkability and bikeability, and how those translate into liveability. These days we aspire to have *complete streets*—those that balance safety and convenience for everyone using the road.

One of the most apparent pedestrian accommodations is the SIDEWALK, a narrow path generally separated from the street. These footpaths can be made from many materials, but they are ribbons of concrete in most cities. Sidewalks may appear simple, but considerable engineering is present in their design and construction. The cracking of concrete is inevitable. Tree roots invade the subsurface, cycles of freeze and thaw lift the soil, and vehicles impose unanticipated loads. Sidewalks are designed with CONTROL JOINTS to constrain the cracks' location to a regular pattern by artificially weakening the concrete.

These INDUCED CRACKS are preferential to the unsightly random arrangement that would otherwise occur. Also, concrete shrinks and expands according to temperature. On small structures, this may be imperceptible, but for long shapes (like sidewalks), the thermal movement can add up. Occasional spaces, called EXPANSION JOINTS, are left in the concrete to prevent sidewalks from buckling or having significant gaps. The joints are typically filled with wood, cork, or RUBBER to allow for movement over time.

Accessibility is the term used to describe how we make sidewalks and other pedestrian facilities safe and efficient for all users, including those with disabilities. Sidewalks have specific minimum widths and slopes to make sure they're not too challenging to traverse. Where a sidewalk meets a curb, it often includes a ramp down to the street's surface. This ramp is called a CURB CUT, and it ensures that users of wheelchairs, walkers, and canes can easily transition onto the path. It also helps pedestrians pushing carts or strollers and even children on bicycles. In addition, sidewalks often feature TACTILE PAVEMENT. These bumpy areas help people with visual disabilities delineate the boundary between sidewalk and road. They are detectable warnings for those who might not otherwise be able to identify a potential hazard, including subway lines, steep grades, stairs, and road crossings. They often have a contrasting color so they're easy to recognize, and many use a familiar texture called TRUNCATED DOMES.

Another critical aspect of pedestrian infrastructure is to allow those on foot to cross streets safely. CROSSWALKS are designated areas for pedestrians to cross a road, making them more visible and predictable to motorists. They are usually located at intersections and marked with large white bars. When a junction has a traffic signal, LIGHTS at either end of each crosswalk show pedestrians when to cross. Some pedestrian signals even include a COUNTDOWN TIMER to show how many seconds are remaining to cross. Depending on the traffic volume, the crossing signal may be simultaneous with the green light for vehicle traffic, or there may be a phase when only pedestrians are allowed to move. Some lights stagger the phases, so pedestrians get a head start from drivers. Some signals operate exclusively on a pre-programmed timer, while others are actuated with a CALL BUTTON at the sidewalk. Even if the switch is present, that doesn't necessarily mean it's connected to the signal controller. Sometimes these buttons are simply *placebos*, or work only during certain times of the day.

Bicycling is one of the most efficient, healthy, and fun ways to get around, but biking in a city without dedicated cycling infrastructure often feels life-threatening. Most places have laws allowing bikes to use the same travel lanes as motor vehicles, but few cyclists feel comfortable doing so except on the least busy streets. There are many different approaches to accommodating bicycle traffic in cities. One of the most straightforward measures is the SHARROW, a symbol used to mark the preferred path for cyclists on shared travel lanes. *Uniformity* is a crucial concept in traffic engineering. If all road users know what to expect, they are less likely to make errors in judgment that lead to collisions. Sharrows don't explicitly provide protection or separation for cyclists, but they help establish expectations between motorists and bicyclists to avoid confusion (and hopefully tension) on the road.

The next step up in bicycling infrastructure is the PAINTED BIKE LANE. These dedicated paths don't provide physical division from vehicles. Still, they visually separate the primary travel lanes, creating a perceptual division between the two streams of traffic (which often have very different prevailing speeds). Bike lines sometimes feature green paint in the United States to further distinguish them from the rest of the road, and they occasionally include painted BUFFERS to create more space between vehicles and cyclists. Separated bike lanes provide the highest level of safety and comfort for those of all abilities. These are exclusive cycling paths with a physical BARRIER of some kind from the main roadway. Of course, separate and dedicated tracks require significant investment, so they are often reserved for only the most highly trafficked routes.

One way to make pedestrians and cyclists safer is to reduce motor vehicles' speed and volume. Changing the posted speed limit is usually not enough to slow down cars, so engineers and city planners employ more creative methods of TRAF-FIC CALMING. At intersections, reducing the CURB RADIUS can slow down vehicles making turns and shorten the crossing distance for pedestrians. However, doing so is feasible only in areas without much truck traffic (since they need more room to turn). Calming options away from intersections include NECKDOWNS to constrict the roadway width, CHICANES to add gentle turns, TREES to reduce the sight distance, and SPEED HUMPS to provide a physical impediment to fast-moving vehicles.

KEEP AN EYE OUT

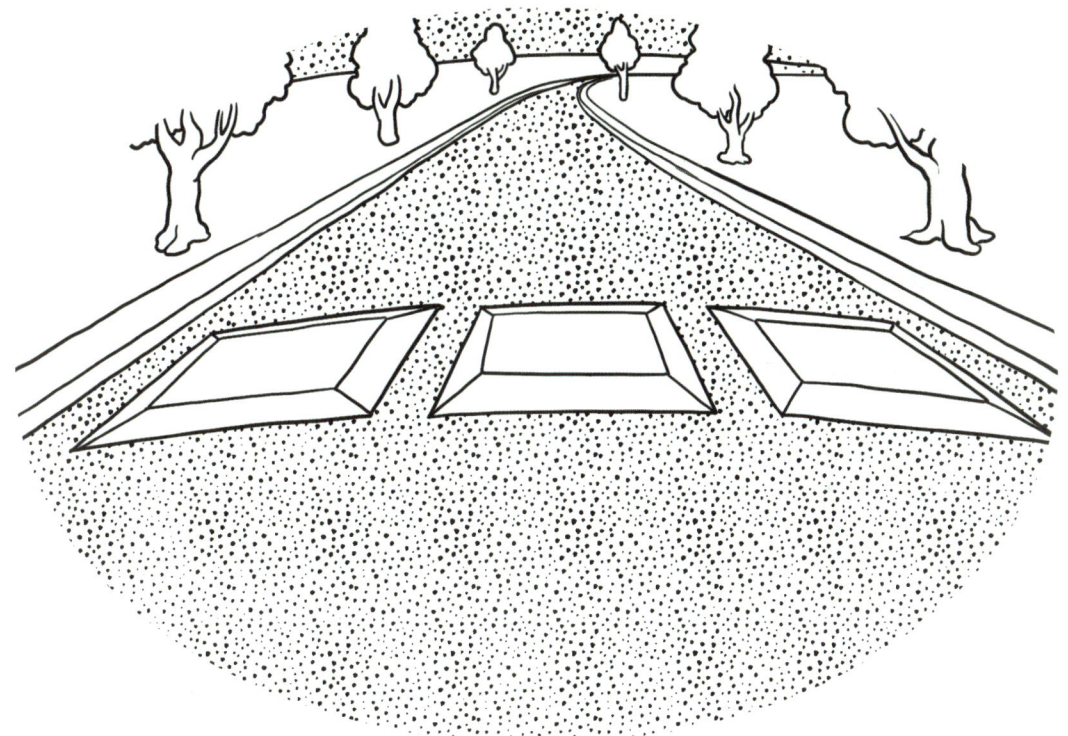

Ever wonder about the difference between speed humps, bumps, and lumps? A *speed hump* is a tool to slow down vehicles on public roadways and is usually 4 meters (12 feet) in width. A *speed bump* is smaller in width but taller in height, and is intended for parking lots and garages. *Speed lumps* (also called cushions) are similar to speed humps, but they have gaps to allow emergency vehicles to pass through without slowing down. These obstacles are unappreciated by motorists, mainly because they are uncomfortable even at the slowest of speeds. Newer designs in development use fluids that harden when drivers go too fast but let slow drivers through without a bump.

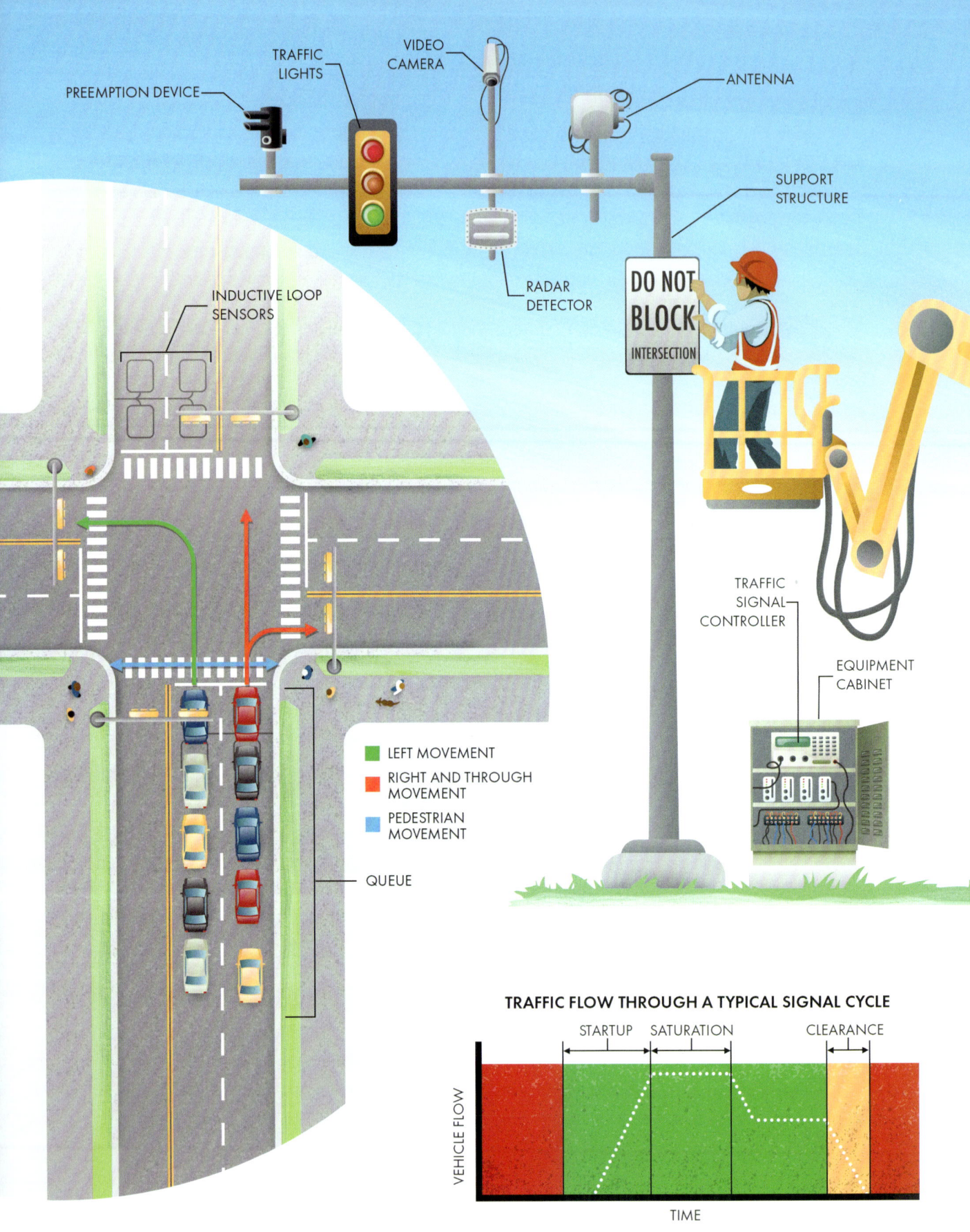

PREEMPTION DEVICE

TRAFFIC LIGHTS

VIDEO CAMERA

ANTENNA

SUPPORT STRUCTURE

RADAR DETECTOR

INDUCTIVE LOOP SENSORS

DO NOT BLOCK INTERSECTION

TRAFFIC SIGNAL CONTROLLER

EQUIPMENT CABINET

LEFT MOVEMENT

RIGHT AND THROUGH MOVEMENT

PEDESTRIAN MOVEMENT

QUEUE

TRAFFIC FLOW THROUGH A TYPICAL SIGNAL CYCLE

STARTUP | SATURATION | CLEARANCE

VEHICLE FLOW

TIME

Traffic Signals

Traffic management in dense urban areas is a complex problem with a host of conflicting goals and challenges. One of the most fundamental of those challenges happens at an intersection where multiple traffic streams—including motor vehicles, bikes, and pedestrians—need to cross one another's paths safely and efficiently. One of the most common ways we control the *right of way* at intersections is the traffic signal. Using signals isn't a panacea for all traffic problems, but they offer a balance of many essential considerations, namely their minimal space requirements and ability to handle large volumes of traffic with only minor interruptions.

Intersections need to be rigidly standardized so that when you come to an unfamiliar one, you already know your role in the careful and chaotic dance of vehicles and pedestrians. That's why almost all traffic signals in a specific area or country look similar. In their simplest form, traffic signals are a set of three LIGHTS facing each lane of an intersection. The lights hang from suspended cables or rigid SUPPORT STRUCTURES. In general, when the light is green, the vehicles in that lane are permitted to cross. When the light is red, they aren't. The amber light warns that the signal is about to change from green to red. Beyond this primary function, traffic signals can take on innumerable complexities to accommodate all kinds of situations.

At each approach to the intersection, vehicles can go in one of three directions, which are called *movements*: RIGHT, THROUGH, or LEFT. Right and through are usually grouped as a single movement, so a typical four-way intersection has two vehicle and one PEDESTRIAN MOVEMENT for each direction. These movements can be grouped into phases of the traffic signal. For example, the left-turn movements on opposite approaches can be grouped into a single phase because they can go simultaneously without conflicts. Traffic engineers determine the grouping of movements into phases and the order of each phase through a signal cycle to accommodate different volumes and types of traffic.

Another critical decision is how long each sequence of a phase should last. Ideally, a green light should last at least long enough to clear the QUEUE built up during the red light, but it isn't always possible, especially during peak times on busy intersections. In cases where the junction is SATURATED, the green light might be extended to reduce the number of cycles since each one includes STARTUP and CLEARANCE times—the periods when the intersection isn't being utilized to its maximum capacity.

The amber light needs to last long enough for drivers to perceive the warning and decelerate their vehicles to a stop at a comfortable rate. Design guidelines consider many factors, but the duration of the amber light is usually set to around one second for every 10 miles per hour or 16 kilometers per hour on the speed limit. In most places in North America, you are allowed to enter an intersection for the entire duration of a yellow light, which means there needs to be a time when all phases have a red light to allow the junction to clear. This clearance interval is usually about one second, but it can be adjusted up or down based on the speed limit and intersection size.

Some traffic signals use a set timing sequence programmed into the controller, but many are more sophisticated than that. *Actuated signal control* is the term we use for traffic lights that can receive input from the outside to adjust timing and phase sequence on the fly. Actuated signals rely on data from traffic-detection systems that can be VIDEO CAMERAS, RADAR DETECTORS, or INDUCTIVE LOOP SENSORS embedded into the road surface. These latter sensors are essentially large metal detectors that can measure whether a car or truck is present (to the annoyance of bicycles, scooters, and motorcycles that are sometimes too small to trigger the loop). Whatever the type of sensor, they all feed data into an EQUIPMENT CABINET located nearby. You've probably seen hundreds of these cabinets without realizing their purpose.

Inside this cabinet is a TRAFFIC SIGNAL CONTROLLER, a simple computer programmed with logic to determine when and how long each phase will last based on the information from the detectors. Actuated control gives a traffic signal much more flexibility to handle variations in traffic load. For example, if a nearby road is closed and traffic is rerouted through an intersection that doesn't usually see such high demand, it may need to be reprogrammed before the closure. A traffic signal equipped with actuated control will simply see the additional traffic and adjust its phasing accordingly. The same is true with special events, like concerts and sports games, that create colossal traffic demands on irregular schedules. Actuated systems can also keep you from waiting at a long light when no one's crossing in the other direction. Finally, actuated control can help prioritize emergency and public transportation vehicles equipped with specialized transmitters. Infrared or acoustic PREEMPTION DEVICES communicate with transmitters on each prioritized vehicle to send the signal controller a call for green.

Actuated control isn't the pinnacle of signal complexity. After all, it still treats each intersection as an isolated entity when it is actually one component of a larger traffic network. Each element of the traffic network can have an impact on other parts of the system. The classic example of this is *gridlock*, where queues of vehicles block adjacent intersections in a way that brings the flow of traffic to a standstill. One solution to this problem is *signal coordination*, where lights can work in synchronization with each other. Coordinated signals are employed on long corridors with small but frequent cross streets. The signals on the major road are timed so that a large group of vehicles, called a *platoon* by traffic engineers, can make it some or all of the way through the corridor without interruption. This coordination can significantly increase the volume of traffic able to pass through intersections, but it works only on stretches of road that don't have other sources of traffic interruptions like driveways and businesses. If the platoon can't stick together, the benefits of coordinating signals are reduced.

The obvious next step in efficiency is coordinating most or all of the signals within a traffic network. This is the job of *adaptive signal control technologies*. In adaptive systems, rather than individual groups of lights, all the information from detectors is fed into a centralized system (often wirelessly with ANTENNAS at each signal) that can use advanced algorithms to optimize traffic flow throughout the city. These systems can dramatically reduce congestion, and many cities have implemented adaptive technologies for their traffic signals.

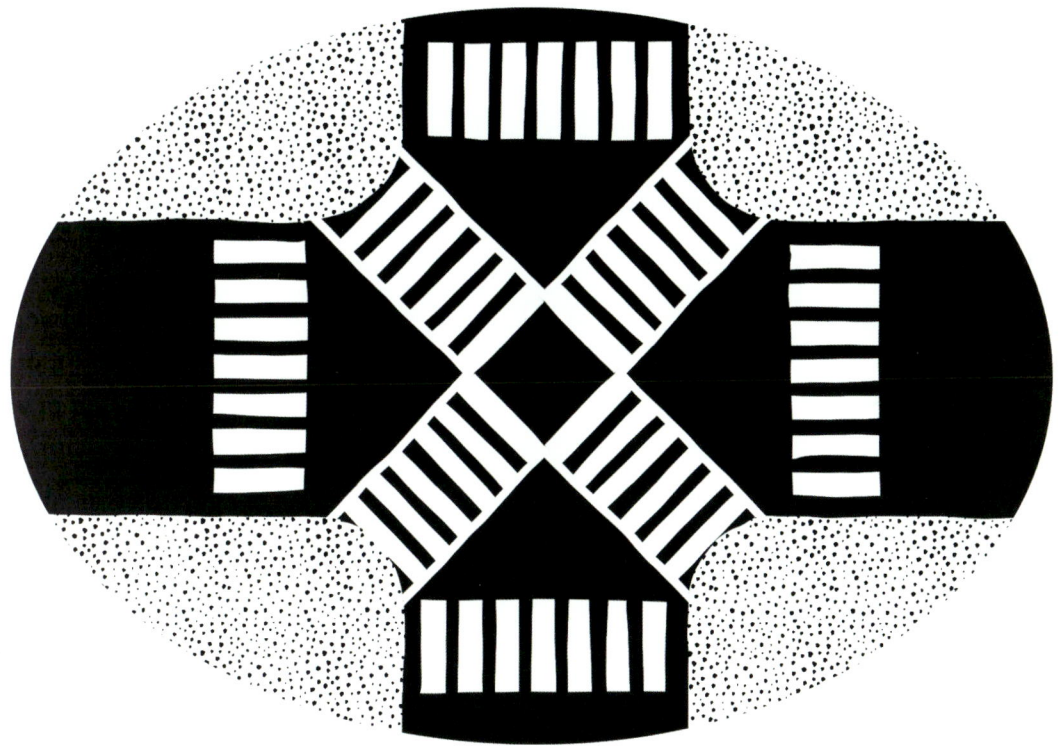

A *pedestrian scramble* is a traffic signal phase that stops all vehicular traffic, allowing pedestrians to cross an intersection in every direction, including diagonally. These scrambles are feasible only at junctions with high volumes of pedestrian traffic since it takes longer to walk across the diagonal, extending the wait time for motorists. They are most common in downtown areas where turning vehicles would have to wait for large numbers of crossing pedestrians if the vehicle and pedestrian movements were simultaneous.

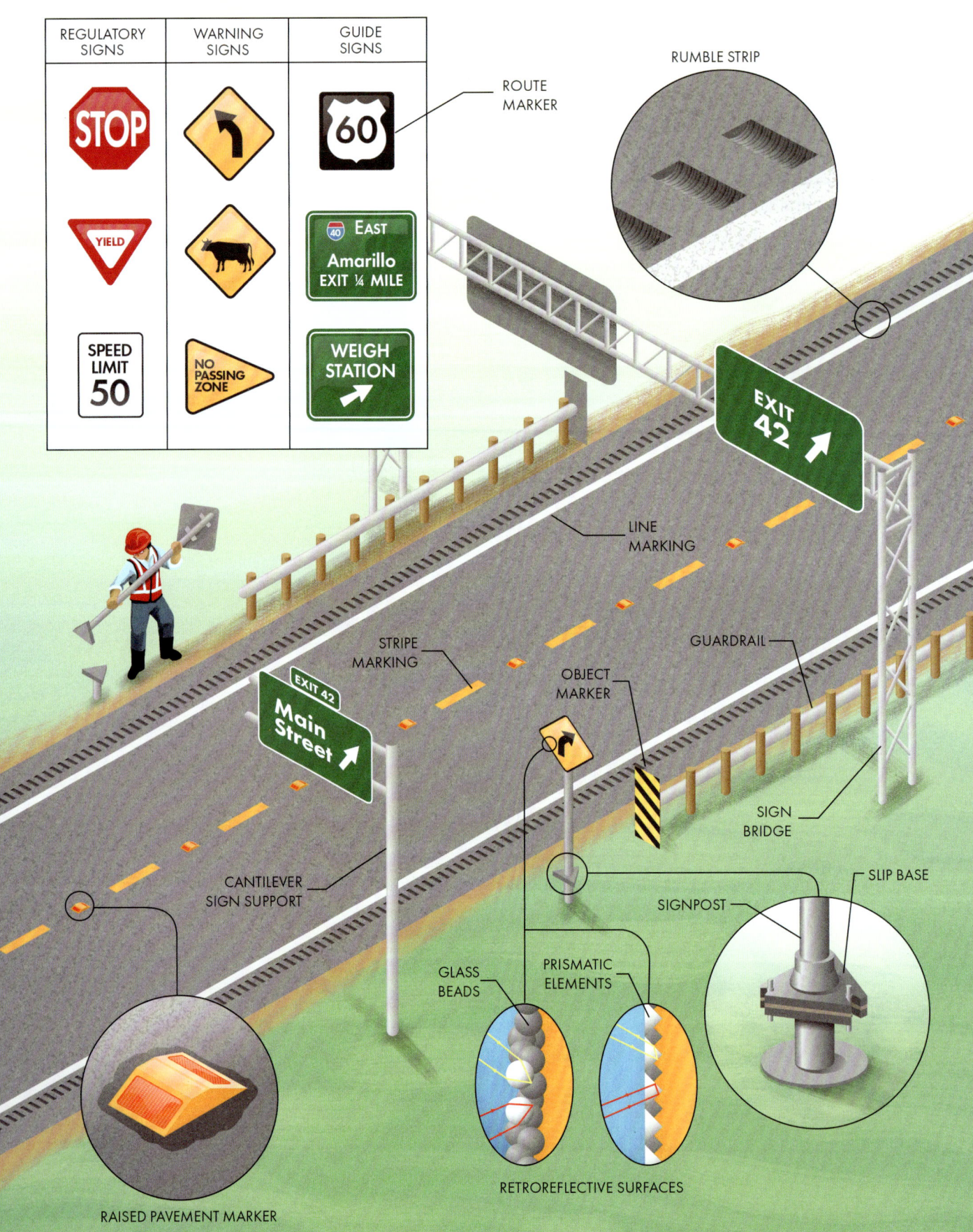

REGULATORY SIGNS	WARNING SIGNS	GUIDE SIGNS
STOP		60
YIELD		I-40 EAST Amarillo EXIT ¼ MILE
SPEED LIMIT 50	NO PASSING ZONE	WEIGH STATION

ROUTE MARKER

RUMBLE STRIP

EXIT 42

LINE MARKING

EXIT 42 Main Street

STRIPE MARKING

OBJECT MARKER

GUARDRAIL

SIGN BRIDGE

CANTILEVER SIGN SUPPORT

SIGNPOST

SLIP BASE

GLASS BEADS

PRISMATIC ELEMENTS

RETROREFLECTIVE SURFACES

RAISED PAVEMENT MARKER

Traffic Signs and Markings

One of the most important aspects of making roadways safe and efficient is the uniformity of signs and markings. Motorists moving at rapid speeds must make snap decisions. When signs are instantly recognizable and understandable, drivers and other roadway users experience less confusion and fewer surprises. That means they are less likely to misjudge a hazard or make a poor decision. The signs and markings used to regulate, warn, or guide traffic on roadways are collectively known as *traffic control devices*. Just about every aspect of their design is rigidly standardized within an individual country (and sometimes internationally). Sizes, shapes, locations, colors, symbols, and words are all carefully prescribed to ensure that drivers will be comfortable and capable of navigating the roadways no matter where they happen to go. It also makes our infrastructure more cost-effective because materials, products, and equipment are standardized across the country. In the United States, the manual that governs the uniformity of traffic control devices is more than 800 pages long. It includes guidelines for just about any situation that could be encountered in roadway design.

Traffic signs need to convey information as clearly and directly as possible because road users have only a moment to recognize, comprehend, and respond to them. Signs are intended to provide their message first by shape, then by color, and finally by meaning or symbol. The most important

signs can be recognized simply by shape (for example, an octagonal stop sign).

Three major categories of signs (along with many minor types) are used on roadways: regulatory, warning, and guide signs. REGULATORY SIGNS inform road users of traffic laws and include speed limit, stop, and yield signs. They mostly use a combination of black, white, and red colors. WARNING SIGNS alert road users to hazards or unexpected conditions. They are almost always yellow diamonds with black writing. OBJECT MARKERS are another type of warning sign to mark obstructions in the roadway or alongside it using diagonal yellow and black stripes. GUIDE SIGNS inform road users of helpful information for navigation and direct them along their way, and they're almost always green with a white border and message. ROUTE MARKERS are another type of guide sign. They use distinctive shapes (often shields) and colors to differentiate road classifications.

Most signs are mounted on metal POSTS adjacent to the roadway. These posts hold the sign high enough to be easily seen by all road users. The other option for sign mounting is overhead structures, which are most common on highways because traffic can obstruct the view of post-mounted signs from the centermost lanes. Overhead sign mounts allow better visibility for all traffic lanes, and they come in two varieties. When carried by only a single vertical member, they are called CANTILEVER SIGN SUPPORTS. These can extend only so far because the load is unbalanced. For wider roadways, the supports are held on both sides by a structure called a SIGN BRIDGE.

Although signs are critical to keeping roadways safe and efficient, they can represent a hazard. Signs' narrow poles can pass through many parts of a car or truck like butter. If an errant vehicle hits a signpost or vertical support, it can significantly worsen the damage and danger of the crash, so signposts are required to be *crashworthy*. In most cases, posts have a *breakaway* feature to reduce the impact on a vehicle if struck, minimizing potential for injury to the occupants. Wooden posts have holes drilled in them, so they easily break when hit. Metal signposts often use breakaway hardware called SLIP BASES. These joints connect using plates with bolts in open slots. When hit, the bolts easily slip out, allowing the signpost to give way. A slip base has the added benefit of streamlining the replacement of a sign that has been knocked over. The concrete and base remain intact, so installing a new sign atop the original base is as simple as bolting it on. Overhead signs can't be designed to break away because a falling sign could endanger other roadway users. Instead, the supports are protected from collisions using a GUARDRAIL, barrier, or crash cushion. (More information about these structures is provided in a later section.)

Another type of traffic control device involves placing markings on the road surface itself. LINES and STRIPES are painted onto the pavement to provide information and guidance to road users. Depending on the traffic levels and budget, different materials are used for these markings, from simple latex paint to *thermoplastics* that are melted onto the road surface. In areas that receive snow, lines are often recessed into the pavement to protect them against snowplows.

RAISED PAVEMENT MARKERS are another surface feature used to guide motorists. They provide both visual and tactile feedback since driving over them makes a noticeable bump. The colors of the reflectors in raised pavement markers carry different meanings. White and yellow are used for marking lanes. Blue markers show the locations of fire hydrants. If you ever see red reflectors, turn around! They are often installed on the backside of pavement markers to warn wrong-way drivers. RUMBLE STRIPS are a type of surface safety device not to be seen but heard. They are created by grinding grooves into the pavement at regular intervals. When a car wanders outside its lane, the sound and vibration from the rumble strip will warn the driver of the deviation.

Traffic control devices aren't much use if they aren't visible in the dark. It used to be common to have dedicated lights to illuminate road signs at night or during poor weather. Now, almost all signs and roadway markings are *retroreflective*, meaning they reflect light back toward its source in the same direction as it came. RETROREFLECTIVE SURFACES take advantage of headlights, bouncing their light directly back toward the vehicle and driver inside. This makes signs and road markings appear much brighter than their non-retroreflective surroundings. Signs are surfaced with plastic sheeting embedded with GLASS BEADS or PRISMATIC ELEMENTS. Retroreflective glass beads are also embedded into the markings on roadway surfaces, making them more visible to vehicles with their headlights on. These beads are sometimes known as *cat's eyes* because they function similarly to the way a cat's eyes appear to glow at night when subjected to light.

KEEP AN EYE OUT

Sometimes a message or warning is so important that it is placed right on the road surface where drivers are sure to see it. However, unlike signs that face directly toward a driver's eyeline, the road's surface can only be seen from a vehicle at a shallow angle. The resulting markings appear foreshortened and challenging to read. And the foreshortening gets even worse when motorists are moving at a rapid speed. As an example, most people significantly underestimate the length of stripes on a roadway. They appear much shorter than the standard 3 meters (10 feet). Letters and symbols on the road surface are elongated to combat this optical illusion and improve the legibility to drivers. In most cases, markings on the roadway surface are stretched two to five times their standard size along the direction of travel. Hold the book at just the right angle to your eyes, and the words in the illustration will look perfectly normal.

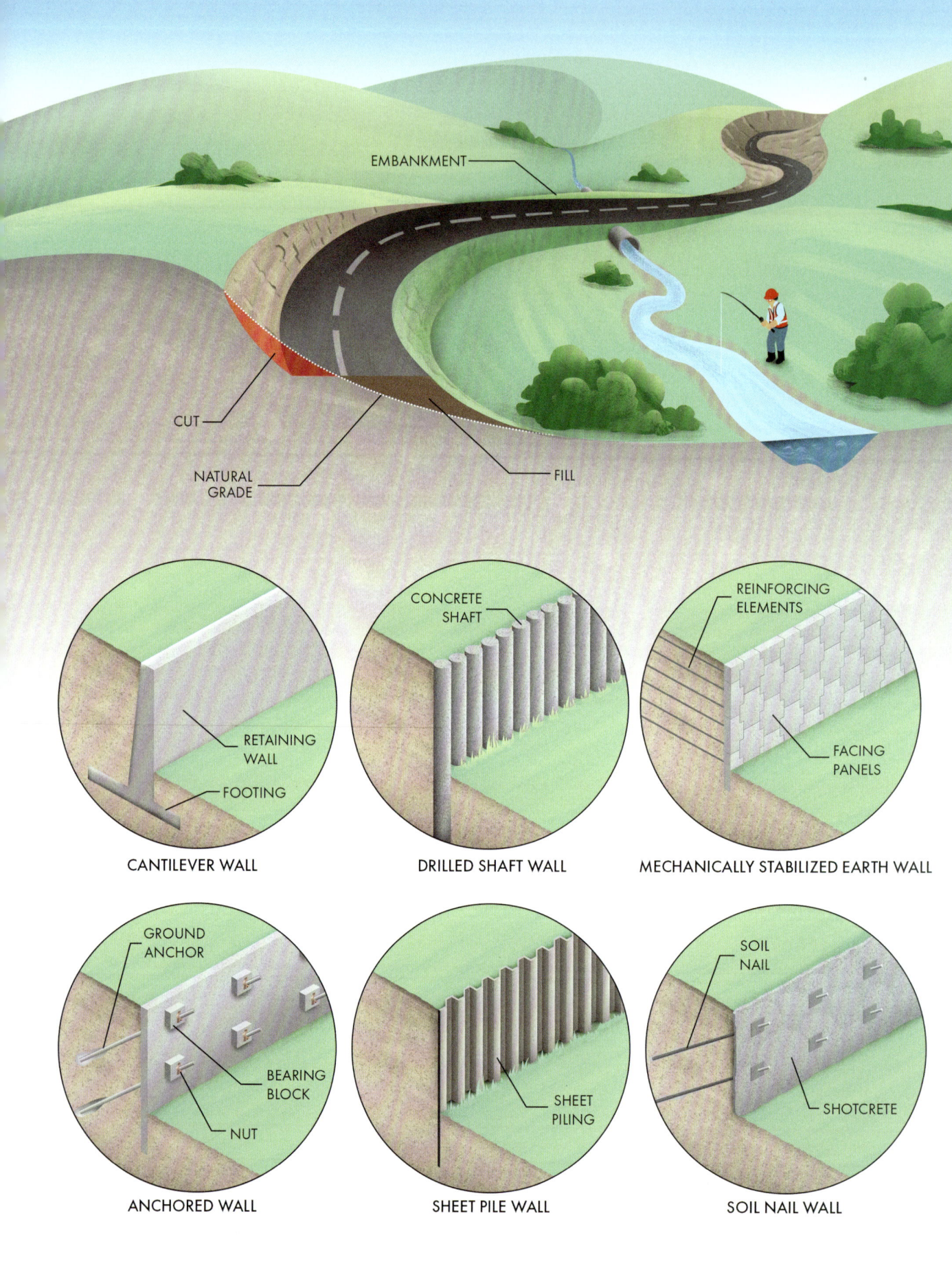

EMBANKMENT

CUT

NATURAL GRADE

FILL

RETAINING WALL

FOOTING

CANTILEVER WALL

CONCRETE SHAFT

DRILLED SHAFT WALL

REINFORCING ELEMENTS

FACING PANELS

MECHANICALLY STABILIZED EARTH WALL

GROUND ANCHOR

BEARING BLOCK

NUT

ANCHORED WALL

SHEET PILING

SHEET PILE WALL

SOIL NAIL

SHOTCRETE

SOIL NAIL WALL

Highway Earthwork and Retaining Walls

The natural landscape is never perfectly suited to roadway construction as it stands. The Earth is just too uneven to traverse at rapid speeds easily. Safe and efficient travel requires gentle curves, both horizontally and vertically. It needs grades that aren't too steep and paths that are relatively direct between points of interest. That means to build a road, we need a way to smooth out the surface of the Earth. All the ways we use to modify the ground's shape and structure are collectively known as *earthwork*, and they can be the most crucial aspect of a roadway construction project.

Engineers and contractors use *cross sections* to communicate the shape of roadways. These drawings show slices through the road along its length, and they are the literal language of road building. On a cross section, you can see the earth's level before construction (called the NATURAL GRADE) and the proposed surface after completion. Any difference in those two lines means some earthwork will be required. Areas above the proposed roadway need to be excavated away, also known as areas of CUT. Excavation is necessary when the final level will be lower than the surrounding terrain, such as through a steep hillside. Areas below the proposed road need to be raised with FILL, such as when passing over a stream or on the approaches to a bridge. Larger areas of fill are often called EMBANKMENTS. Cuts and fills are the most fundamental elements of any earthwork project. Of course,

you can't visually compare before and after the earthwork is accomplished side by side, but it is often apparent where the natural landscape has been modified once you start paying attention.

You might notice that cuts and fills usually tie into the natural grade on a slope. That's because a soil's strength almost entirely depends on internal friction between the individual soil particles. Pour out some sand on a table, and you'll notice that the pile doesn't stand straight up. Instead, it forms a slope. This slope's angle is called the *angle of repose*, which is the steepest angle at which a soil can naturally rest. Add some weight to the top of your pile, and it will collapse even further.

A slope's stability can vary significantly depending on the type of soil and the loading it needs to withstand, but engineers rarely trust anything steeper than around 25 degrees. That means a constructed slope must be at least twice as wide as it is tall, which can be a problem for two reasons. First, it takes about twice as much material as a slope that can stand vertically, requiring a lot more excavation or fill to construct. Second, it takes up more space, which can be at a premium, especially in crowded cities. In many situations, it makes sense to avoid these disadvantages by using a RETAINING WALL to support a steep (and even vertical) slope.

Soil doesn't flow as easily as water, but it is around twice as heavy. Thus, the force exerted on a retaining wall, called the *lateral earth pressure*, can be enormous. Accordingly, retaining walls must be quite strong to withstand this pressure. Many different types of retaining walls solve this problem in various ways. You'll notice a variety of these walls in the constructed environment if you know where to look. They aren't only for roadway projects, but that's a common application. The most basic retaining walls rely on gravity for their stability, often employing a FOOTING to create a CANTILEVER WALL. In this configuration, the wall can use the weight of the restrained soil to its advantage. The earth sits atop the footing, which acts as a lever, helping keep the wall upright against the lateral forces.

Some retaining walls use GROUND ANCHORS (also called *tiebacks*) to provide horizontal stability. The anchors consist of steel strands or bars grouted into the soil behind the wall. Once installed, a hydraulic jack applies tension to each anchor, and tapered wedges or NUTS lock the anchors firmly against the wall. BEARING BLOCKS or plates are often used to distribute the anchor load across a larger area, recognizable from the outside from their repeating pattern.

Another kind of retaining wall uses *piles*, vertical members driven or drilled into the ground. They include reinforced CONCRETE SHAFTS installed with a drilling rig like gigantic fence posts. They also include interlocking steel shapes called SHEET PILING. Pile walls are often used for temporary excavations during construction projects because the wall can be installed first before digging begins, ensuring that the excavated faces have support for the entirety of construction.

One common type of retaining wall involves tying a mass of soil together to act as its own wall. This can be accomplished during the fill operation by layering reinforcement elements between each lift, a technique called MECHANICALLY STABILIZED EARTH. The REINFORCING ELEMENTS can be steel strips or fabric made from plastic fibers called *geotextile* or *geogrid*. When the natural ground is excavated to create a steep face, adding layers of reinforcement isn't feasible. Instead, SOIL NAILS can be inserted into the slope as reinforcement. Like ground anchors, soil nails consist of steel rods grouted into drilled holes. But unlike anchors, they are not tensioned. Instead of applying a force to the wall's face, their job is to secure the soil mass together to support itself and the soil behind it.

Both mechanically stabilized earth and soil nail retaining walls use concrete on the outside face of the wall. These facings are rarely supporting much of the load. Instead, their job is to protect the exposed soil from erosion, and in permanent applications, improve the wall's appearance. In temporary situations, the facing sometimes consists of SHOTCRETE, a type of concrete that can be sprayed from a hose using compressed air. For permanent installations, they often use interlocking concrete PANELS with a decorative pattern. These panels not only look nice, but they also allow for some movement over time and for water to drain through the joints.

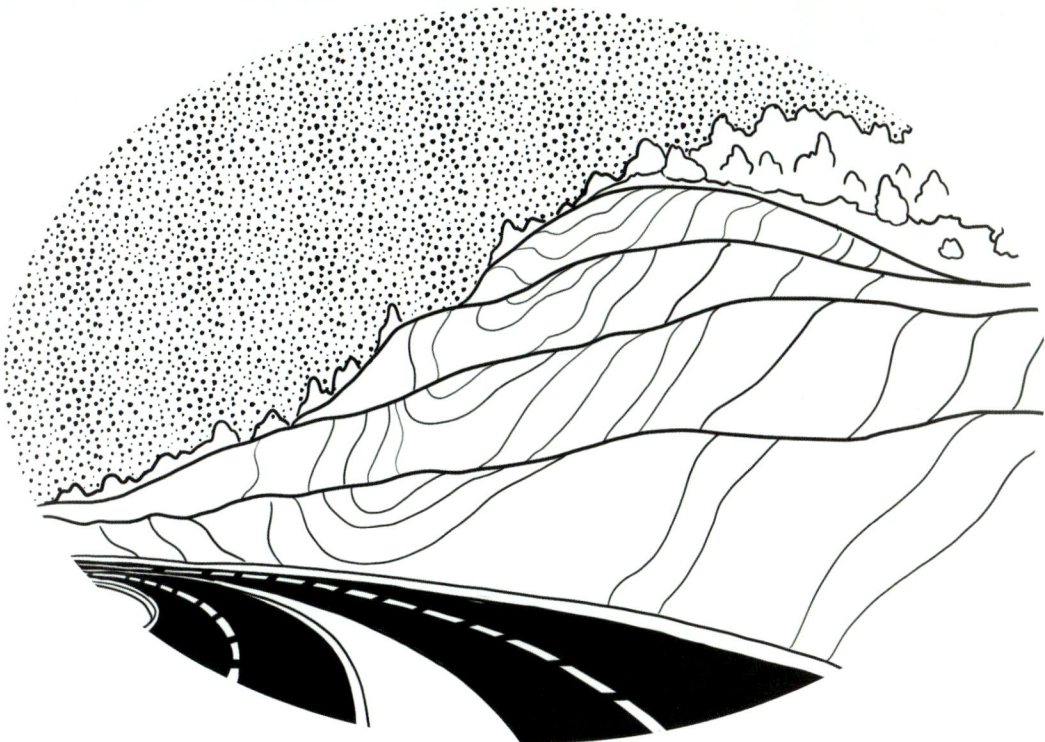

Sometimes a road cut is made where the underlying ground is mostly rock rather than soil. It is much more difficult to excavate through rock than soil, but a retaining wall usually isn't required to support the exposed faces (since rock can often be trusted—after a detailed engineering analysis—to hold itself up). That means many road cuts are left entirely uncovered, revealing extraordinary snapshots into the surface of the Earth. These road cuts may appear as boring walls of rock at first glance. But to a geologist, their strata are indispensable glimpses into how different landscapes were formed. In fact, roadside geology is a hobby on its own, with guidebooks available for many parts of the world. From chalky limestone to swirling marble, you can gain a whole new appreciation of our rocky planet from the comfort of your vehicle. Be careful, though, because this hobby can be both a figurative and literal slippery slope! You may find yourself planning routes on road trips based on which rocks are visible via the highway, but always be sure to take precautions when stopping near busy roads and when clambering over steep terrain.

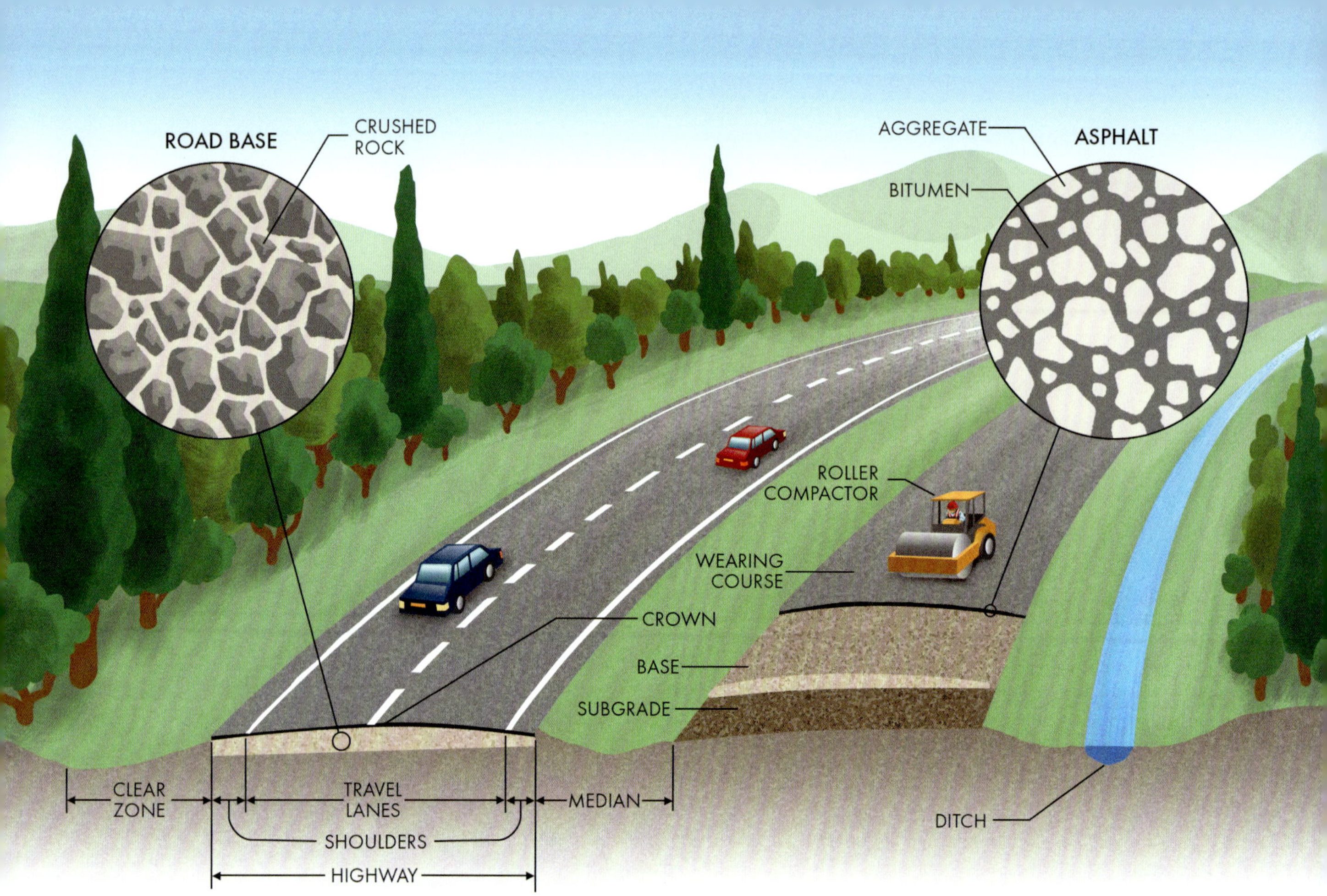

ROAD BASE
CRUSHED ROCK

AGGREGATE
BITUMEN
ASPHALT

ROLLER COMPACTOR

WEARING COURSE

CROWN

BASE

SUBGRADE

DITCH

CLEAR ZONE

TRAVEL LANES

SHOULDERS

MEDIAN

HIGHWAY

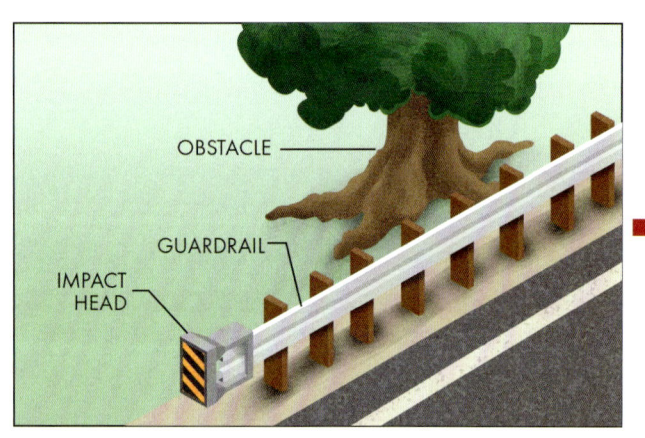

OBSTACLE

GUARDRAIL

IMPACT HEAD

DEFORMED RAIL AFTER IMPACT

CRASH CUSHION

JERSEY BARRIER

Typical Highway Section

I'm often asked why road construction projects seem to take so long when the finished product is just a simple ribbon of pavement sitting on the ground. It's not because of idle construction workers or dishonest contractors. It's because highways are complicated. Ensuring roadways can carry the amount and weight of modern cars and trucks and allow them to travel safely at such incredible speeds is no small feat. The only reason it seems ordinary to us is that roadways are so carefully designed and constructed. From the ground up, highways have many features to make fast and efficient vehicle travel possible.

You only see the outer surface when driving, but there is much more to a roadway structure below the surface. Roads are built in layers, sometimes called *courses*, to make them durable and long-lasting. Before any new roadway can be installed, some earthwork is required to smooth the surface of the earth (as described in the previous section). The layer of existing soil upon which a road is constructed is called the SUBGRADE, and it's not always well-suited to withstand the tremendous and frequent loads from vehicular traffic. Instead, one or more courses of ROAD BASE are placed and compacted on top of the subgrade, often made from crushed rock. Road base serves a variety of purposes. It provides a stable platform during construction, distributes the weight of vehicles evenly to the subgrade, provides drainage for any water that infiltrates below the road, and protects the pavement from frost.

The top layer of pavement is the WEARING COURSE because it's the one exposed to the controlled chaos of constant motor vehicle traffic. Concrete is occasionally used as the wearing course for major highways because it is exceptionally hard and durable. Concrete consists of cement, rocks (known as AGGREGATE in the industry), and water, and it can withstand huge volumes of heavy truck traffic better than any other pavement. But concrete has some disadvantages, too. It is expensive to install. It's hard to repair because it takes a long time to cure, extending the duration of road and lane closures. And it can also be too slick when wet, so it must be grooved for traction with tires. That's why, instead of concrete, most roadways are paved using ASPHALT.

Asphalt pavement has only two primary ingredients: aggregate and BITUMEN, a thick, sticky binder material from the refinement of crude oil. Asphalt just ticks so many of the boxes needed for modern roadways. The materials are readily available. It provides excellent traction with tires without needing grooves. It's flexible, so it can accommodate some movement of the subgrade without failure. Finally, it's easy to repair. Asphalt is heated into a workable mix, placed atop the base course, and then compacted into place with heavy ROLLERS. It is ready for traffic almost as soon as it cools down.

The term HIGHWAY is usually used to describe the entire width of the paved road. It consists of the TRAVEL LANES in which vehicles drive and SHOULDERS, which serve as emergency stopping lanes for vehicles that break down. Shoulders are usually narrower than travel lanes and occasionally paved to a lesser thickness to save cost, so they can't accommodate regular travel. Although highways might look flat, they usually slope toward the edges, giving the center of the road a CROWN. A flat surface doesn't shed water quickly. Any accumulation of water is dangerous to vehicles by making roads slippery and creating more ice in the winter. Crowning the roadway accelerates the drainage of precipitation and keeps the surface of the road dry. Once the water has reached the edge of the pavement, it needs somewhere to go. Otherwise, it can soften and weaken the soils underlying the road. Highways often include DITCHES along the roadside to carry rainwater away. (See Chapter 7 for more details on drainage structures.)

Some of the most dangerous crashes occur when a vehicle veers off the road due to a hazard or loss of control. Many safety features on highways are designed to prevent roadway departures from turning into serious collisions. Major routes often separate the two traffic directions with a MEDIAN, creating a divided highway. The median between the roadways is a grassed area used to prevent errant vehicles from crossing into the opposing traffic, reducing head-on collisions. Most highways also include a CLEAR ZONE along the outside

of each highway, an unobstructed area that gives drivers room to stop or regain control if their vehicle leaves the roadway. The clear zone is kept free of obstacles like trees, signs, and utility poles that could make a crash more serious. When signs must be placed in this area, they have breakaway supports to reduce the impact of a potential collision. And when an obstacle in the clear zone can't be removed or made crashworthy, it must be protected with a barrier.

Longitudinal barriers keep vehicles from leaving the roadway when a dangerous OBSTACLE or drop-off is present. They are also used in place of or in addition to a median between a divided highway. There are many types of barriers used for various situations, and they all go through full-scale crash testing before being used on active roadways. Steel GUARDRAILS can deflect when hit, which somewhat softens the blow of a collision but also means they must be replaced after each crash. Another common type of longitudinal barrier, called a JERSEY BARRIER, is made of concrete. Its shape allows a tire to ride up the side, often redirecting the vehicle without causing significant damage.

One challenge with longitudinal barriers is that their blunt ends can create a dangerous obstacle in the clear zone. Most feature end treatments to lessen the severity of a collision if struck. Steel guardrails often feature an IMPACT HEAD that slides along the rail when hit, DEFORMING it to absorb the energy of the crash while redirecting it to the side to protect the vehicle's occupants. Rigid barriers often include a CRASH CUSHION at their ends. There are a wide variety of designs, but the most common feature barrels filled with sand or crushable steel components that can absorb the energy of a collision, significantly reducing the severity.

Unlike concrete, asphalt pavement doesn't go through a chemical reaction to cure. Instead, we use temperature to transform it from a workable mix to a stable driving surface, a process that is entirely reversible and repeatable. That means asphalt is nearly 100 percent recyclable. In fact, asphalt concrete is one of the world's most recycled materials by weight. Many of the roads you drive on every day probably came, at least in part, from other nearby streets or highways that reached the end of their life. We even have equipment that can recycle pavement in place, minimizing traffic interruptions and the costs of hauling all that material to and from the job site. A typical paving "train" consists of a milling machine to remove the old asphalt, a recycling unit to warm and mix it with additives, a paver to place the rejuvenated asphalt, and compactors to compress it into place.

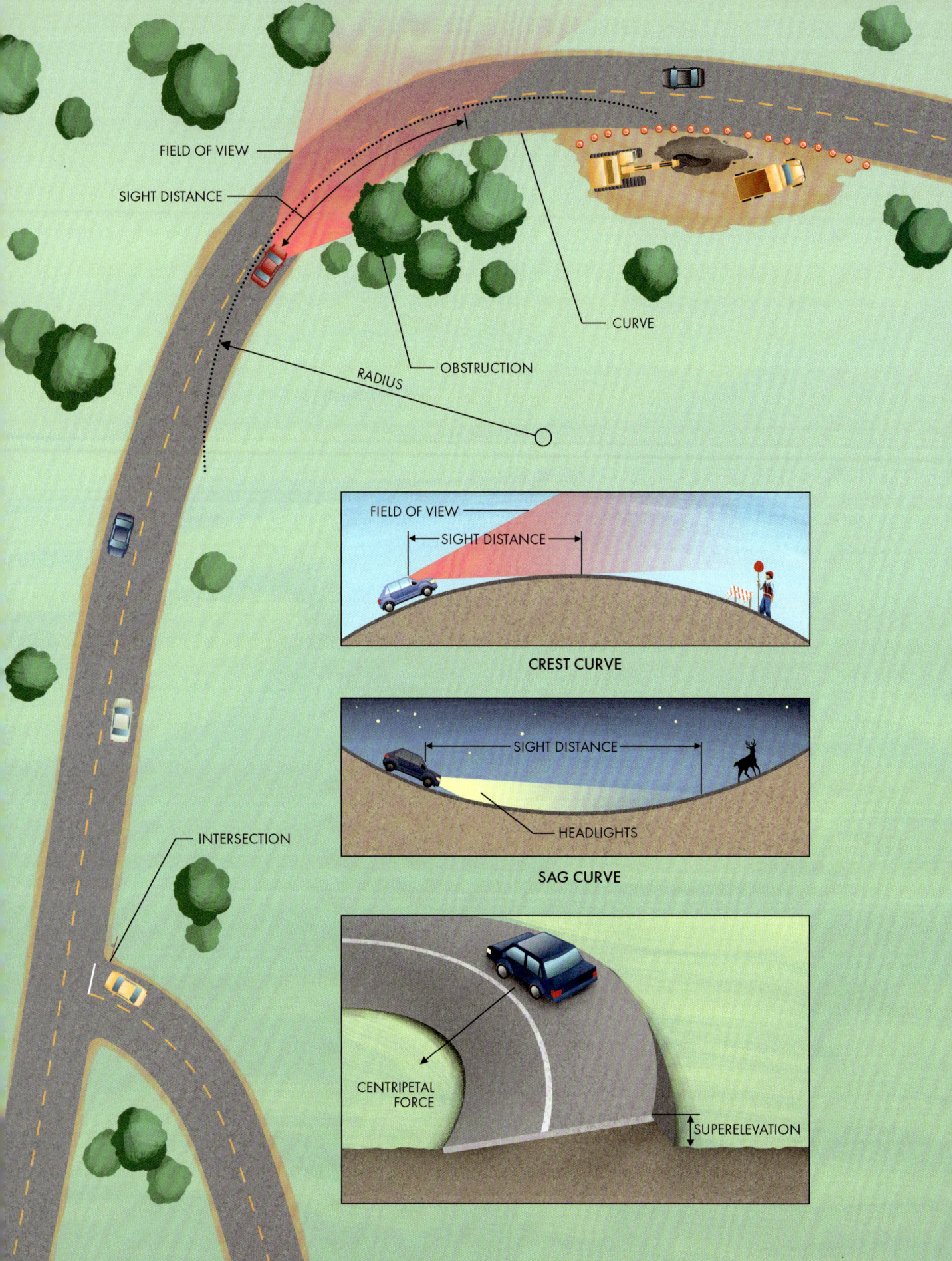

FIELD OF VIEW

SIGHT DISTANCE

CURVE

OBSTRUCTION

RADIUS

INTERSECTION

FIELD OF VIEW

SIGHT DISTANCE

CREST CURVE

SIGHT DISTANCE

HEADLIGHTS

SAG CURVE

CENTRIPETAL
FORCE

SUPERELEVATION

Typical Highway Layout

One class of roadway looks much different from urban arterial and collector roads. I use the term *highway* in this book, but others may call them freeways, motorways, expressways, or throughways. Whatever way you call them, they achieve the pinnacle of traffic capacity through *controlled access*. For smaller highways, that means a reduced number of driveways and at-grade INTERSECTIONS. For the highest capacity roads, it means the only way to enter or exit them is by a ramp or interchange (more on those in a later section). Controlling access to the road reduces interruptions, allowing a relatively unhindered flow of high-speed traffic. That increased speed usually means increased roadway capacity. However, it also decreases the time for decision-making by drivers and, thus, increases the potential for dangerous crashes. It is remarkable that we're able to put ourselves in metal boxes that hurtle away at incredible speeds from place to place, and highways include a multitude of safety features that make such travel possible. This safety starts at the most basic level possible in how the roadway unfolds as you drive along (often simply called its *layout*).

In an ideal world, every road would be a straight flat path, and we could race along at whatever speed we wanted. But all highways include hazards like curves, hills, traffic, obstacles, and weather. Reality dictates that we balance vehicles' speed with drivers' abilities to navigate such dangers. Three essential speeds on highways are not always equal: the *design speed*, the posted *speed limit*, and the speed at which any individual driver chooses to travel. Drivers select their speed based on their personal skill level, comfort, and perception of hazards. Roadway operators set the speed limit based on widely accepted standards for safety. Highway designers choose a design speed to ensure that all the geometric features along the roadway are consistent and appropriate for the eventual velocity at which most drivers will travel.

A highway's *alignment* is its horizontal layout—the way it appears when looking from above. All roads include CURVES needed to change the direction of travel, and those curves can present severe challenges to drivers if not designed correctly. Any object changing its direction needs a CENTRIPETAL FORCE toward the center of the turn. Otherwise, it will just continue in a straight line. When you feel pushed to one side of the vehicle during a turn, this is your body's inertia trying to keep you moving straight while your car turns. For a vehicle, the centripetal force comes from the friction between the tires and the road. This force increases as the RADIUS of the turn decreases. At a certain speed and turning radius, the required centripetal force can exceed the tire friction, leading a vehicle to skid off the roadway. To avoid this dangerous condition, engineers select the minimum turning radius of curves depending on the roadway's design speed— the faster the rate, the gentler the curve.

Rubber tires provide traction against a road surface, but we can also use geometry to make curves safer for drivers. Highway designers often make the outside edge higher or SUPERELEVATED above the centerline to reduce the need for tire friction around curves. Banking a roadway around a turn takes advantage of the *normal* (that is, perpendicular) *force* from the pavement to provide some or all of the needed centripetal force. In general, the faster the design speed of the road, the steeper the bank around the bend. Superelevation also makes travel around curves more comfortable because the centrifugal tendency pushes passengers into their seats rather than out of them. If the superelevation angle is just right and you're traveling at precisely the roadway's design speed, the level in your cup of coffee won't change at all around the bend.

One other important aspect when designing a horizontal curve comes from the simple but crucial fact that drivers need to see what's coming up to react accordingly. SIGHT DISTANCE is the length of the roadway visible to the driver at any given point in time. On a straight and level section of highway, it is limited only by the driver's visual acuity. However, any time a roadway changes direction, a driver's FIELD OF VIEW can be blocked by OBSTRUCTIONS. If the sight distance isn't enough to recognize and respond to a hazard, a crash may result. The faster you're traveling, the more distance you need to observe turns or obstacles and decide how to manage them. Even if a curve is gentle enough for a car to traverse without skidding, it may

not have enough sight distance for safety due to an obstacle like a hill or wooded area obscuring the driver's view. In such a case, the highway designer would need to increase the curve's radius to lengthen the driver's sight distance to a safer extent (or just remove the obstacle).

The final aspect of roadway geometry is the vertical configuration, also known as the *profile*. Roads rarely traverse perfectly flat areas. Instead, they go up and over hills and down into valleys. The slope, or *grade*, of a roadway is an important design decision. Roads that are too steep make travel difficult, especially for heavy trucks. Uphill stretches are slow-going, and long downhill sections can overheat vehicles' brakes. Grade changes also must happen smoothly to avoid bumps and jerks for the comfort of drivers. On top of all that, vertical curves have the potential to reduce drivers' sight distance.

CREST CURVES—the ones that are convex upward—cause the roadway to hide itself beyond the top. If you're traveling quickly up a hill, a stalled vehicle or animal on the other side could take you by surprise. A crest curve that is too tight won't give you enough sight distance to recognize and react to the obstacle. So, designers must make sure that these curves are sufficiently gentle so that you can still see enough of the roadway as you go up and over. SAG CURVES—the ones that are concave upward—don't have this same issue. During the day, you can see all the roadway on both sides of the curve. However, at night things change. Vehicles rely on HEADLIGHTS to illuminate the road ahead, and sometimes this can be the limiting factor for sight distance. If a sag curve is too tight, headlights won't throw as far. The result is that the sight distance is reduced, making it difficult to react to obstacles at night.

KEEP AN EYE OUT

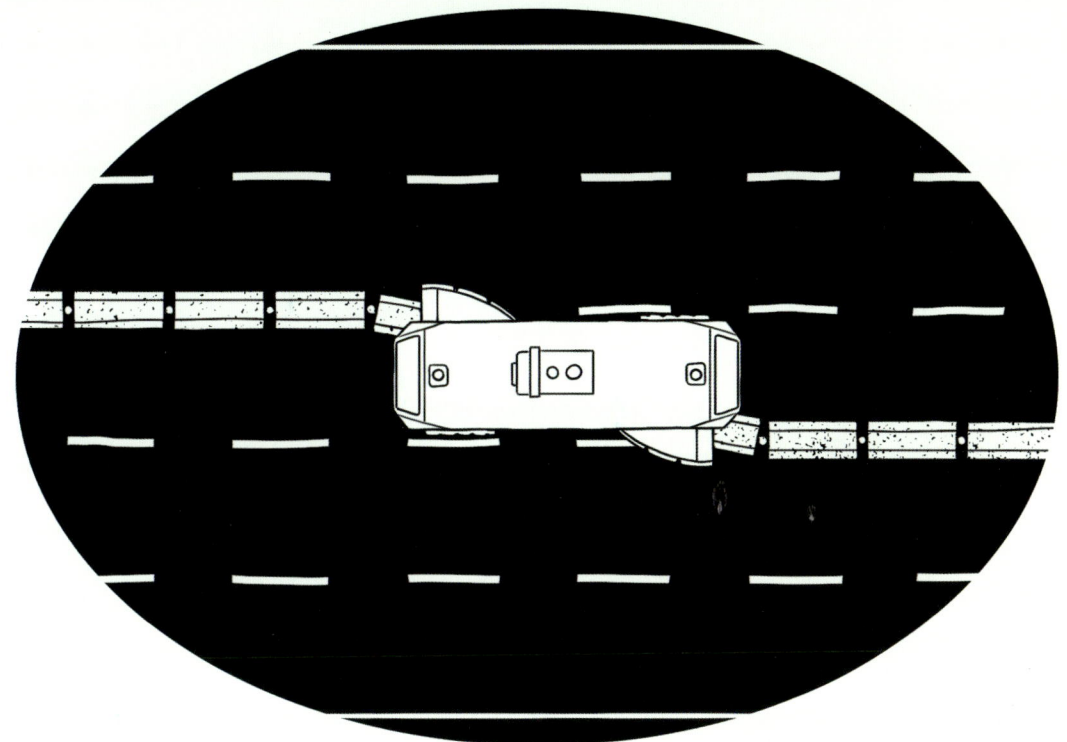

Although they are both called *rush hour*, the peak periods of traffic during the morning and evening commutes are never identical. In large metropolitan areas, it is typical to have more vehicles traveling toward the city center in the morning and away from it during the evening. This tidal flow of traffic often leads to roadways being underutilized, with heavy congestion in one direction and light traffic in the other. It's certainly frustrating to be stuck in traffic with so much empty pavement alongside you. Many places take advantage of those unoccupied lanes by making them reversible, so their travel direction depends on the time of day. There are many ways to achieve such bidirectionality, but one of the most effective is the moveable barrier. Certain roads are equipped with hinged concrete dividers that can be shifted between lanes. Twice per day in the lulls between morning and evening, a machine traverses the roadway, "zippering" the barriers to reverse one or more lanes' direction and increase the capacity during each rush hour.

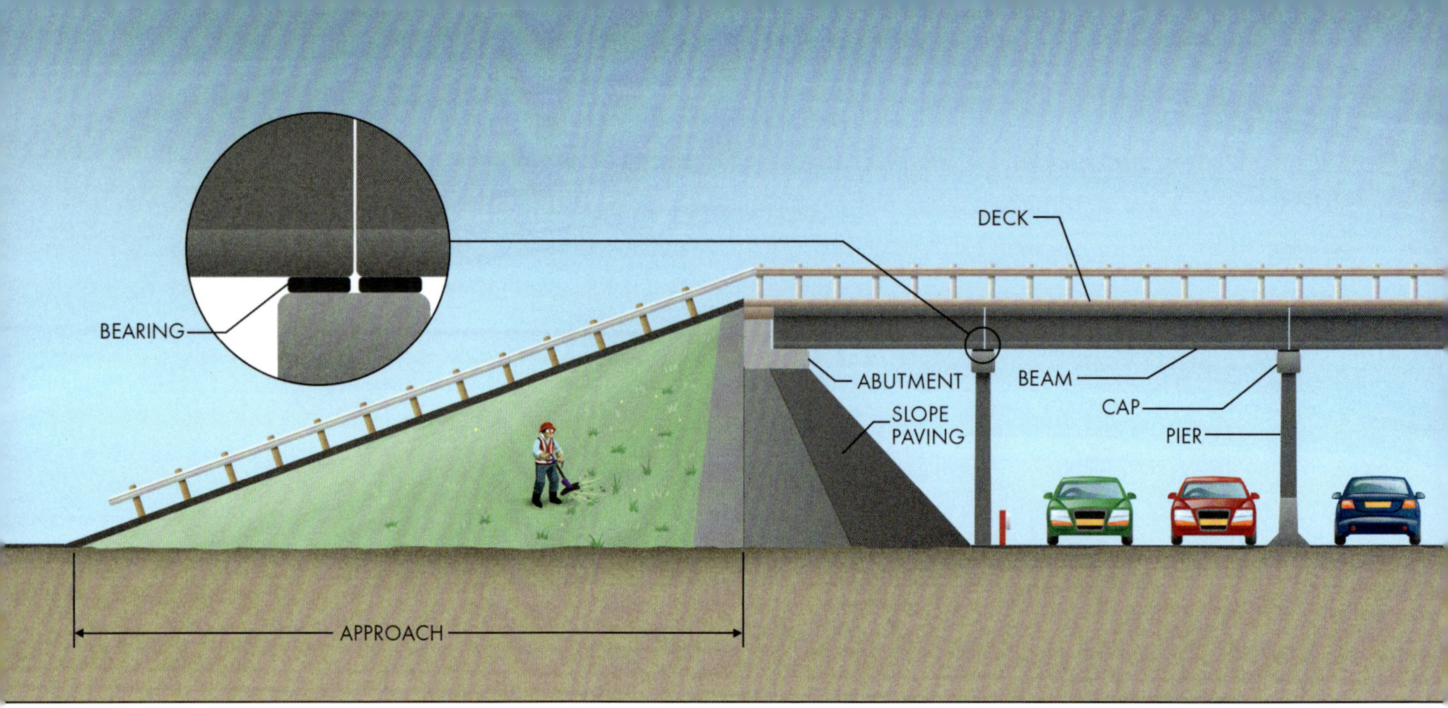

BEARING

DECK

ABUTMENT

SLOPE
PAVING

BEAM

CAP

PIER

APPROACH

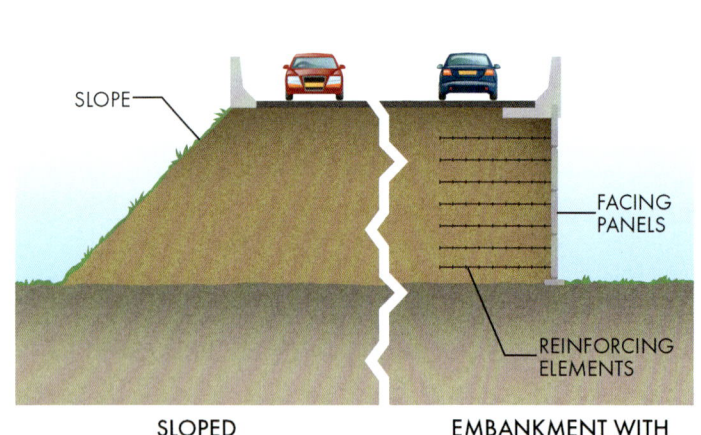

SLOPE

FACING
PANELS

REINFORCING
ELEMENTS

SLOPED
EMBANKMENT

EMBANKMENT WITH
RETAINING WALL

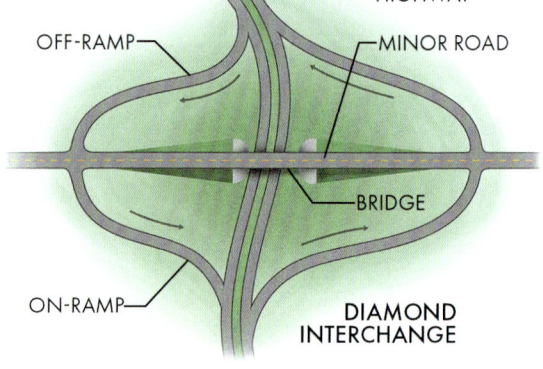

OFF-RAMP

HIGHWAY

MINOR ROAD

BRIDGE

ON-RAMP

DIAMOND
INTERCHANGE

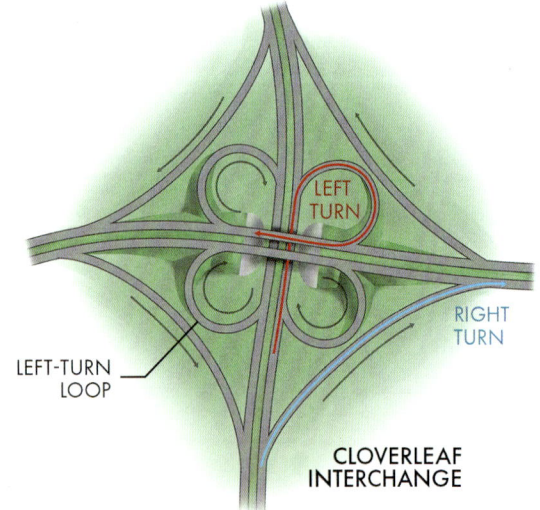

LEFT
TURN

RIGHT
TURN

LEFT-TURN
LOOP

CLOVERLEAF
INTERCHANGE

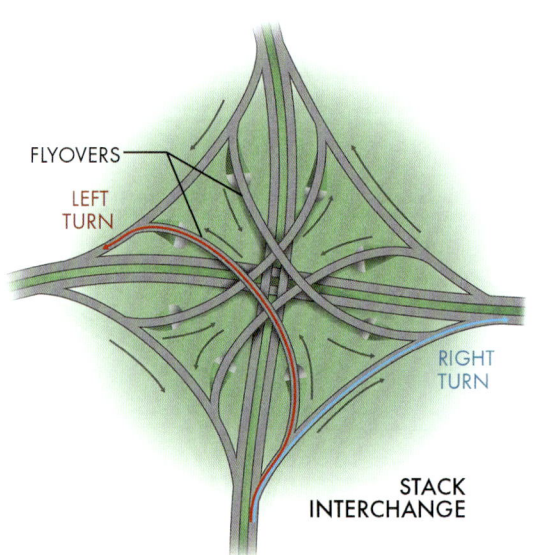

FLYOVERS

LEFT
TURN

RIGHT
TURN

STACK
INTERCHANGE

Interchanges

As discussed in previous sections, when roads intersect, it almost always creates a challenge. Multiple streams of traffic need a way to occupy the overlapping space safely. When the intersection is at grade (in other words, on the ground), traffic flow must be interrupted. Through signs, signals, or roundabouts, the *right of way* is assigned to each traffic stream individually while the others wait. Those frequent starts and stops aren't desirable on a highway that uses controlled access to reduce interruptions, allowing a relatively unhindered flow of high-speed traffic. Instead, entrances, exits, and intersections of highways are often accomplished through grade-separated junctions, also known as interchanges. Grade separation allows streams of traffic to cross each other safely and efficiently without interruption.

One of the most common types of grade-separated junction is the DIAMOND INTERCHANGE, used where a controlled-access HIGHWAY crosses a MINOR ROAD. Off-ramps diverge from the highway, meeting the minor road at a right angle. The OFF-RAMPS become ON-RAMPS past the minor road, returning to the highway. The two conventional intersections formed at the ramps are controlled via signs or traffic signals. One of the roadways will include a BRIDGE, also called an overpass, to achieve the grade separation. Highway bridges have many features to be observed from the outside.

The bridge's *superstructure* includes the BEAMS, structural members that support the DECK upon which vehicles drive. The bridge's weight and all the cars and trucks on top must be transferred to the bridge's foundation. This is accomplished by the *substructure*. ABUTMENTS provide support to the beams at each end of the bridge, accommodating the horizontal and vertical loads of the superstructure. The intermediate supports between each of the bridge's spans are called PIERS when consisting of a single column and *bents* when a frame of multiple columns is used. They are usually designed to handle only vertical loads to be simpler and smaller than the two abutments. In some cases, the piers include CAPS to distribute the forces across each beam and column evenly.

Bridges may seem like static structures, but they must be somewhat flexible. Vibrations from vehicles, settlement of the foundation, expansion and contraction due to temperature, and even forces from the wind can introduce small movements in the superstructure. Rather than make the bridge stiff enough to withstand even the tiniest motion, most bridges use BEARINGS to accommodate this movement, often made from layers of rubber and steel. These pads can transfer the loads of the bridge while still allowing some movement of the superstructure.

The transition between an at-grade roadway and a bridge is called the APPROACH, and it usually consists of an earthen EMBANKMENT. Soil is compacted in layers to create a smooth path up to the bridge. Earth isn't stable with a vertical face, so embankments often have SLOPES on each side. The slopes are generally covered in grass to protect against soil erosion. However, grass doesn't grow well in the shady areas below a bridge. Concrete slabs, called SLOPE PAVING, are often installed on the sloped soil faces below a bridge as armoring against erosion. (Chapter 4 includes further details about bridges.)

One problem with a sloped embankment is how much space it takes up. In urban areas, approach embankments frequently rely on RETAINING WALLS for support, freeing up valuable space. These retaining walls often consist of REINFORCING ELEMENTS layered into the embankment with interlocking concrete FACING PANELS, a technique called *mechanically stabilized earth*. (See the previous section for more information about retaining walls.)

Interchanges become more complicated when two or more highways meet. The ideal junction allows each traffic stream to transition onto any crossing roadway in any direction without interruption. There are many ways to accomplish such connectivity, each with its own advantages and disadvantages. One of the most basic types is the CLOVERLEAF INTERCHANGE, named for its distinctive shape on a map. In a cloverleaf interchange, the vehicles turning RIGHT follow a gentle curve to transition to the crossing roadway. The LEFT turners go past the intersection, then follow a sharp right-hand LOOP onto the

other route in the opposite direction. Cloverleaf interchanges require only a single bridge, so they are relatively inexpensive to construct. However, they have some disadvantages too. Most significantly, the left-turn entrance ramps come before the exit ramps, forcing traffic entering and leaving the highway to weave across each other. This weaving can significantly limit the interchange's capacity.

Another type of grade-separated junction is the STACK INTERCHANGE. In this type of junction, the RIGHT turns usually remain at grade, just like a cloverleaf. However, the LEFT turns are handled by elevated ramps, often called FLYOVERS. The two pairs of left-turning ramps must be stacked above or below the highways, giving this interchange its name. Stack interchanges generally have the highest capacity of all different types of four-way junctions. However, they are complex and expensive structures due to the many layers of elevated roadways required.

Many other types of highway interchanges exist, and most junctions found in the real world borrow elements from various designs. Urban areas put many constraints on such massive structures, including the number, size, and direction of roads connected and the available space for all those ramps (not to mention the two ever-present constraints on all infrastructure projects: schedule and budget). The largest and most complex interchanges are often referred to as *spaghetti junctions*, towering tangles of interweaving ramps transitioning traffic in every direction. On road trips, I am guilty of devising my routes to make sure I pass over each interchange's very top level to get the best (if momentary) views of the city.

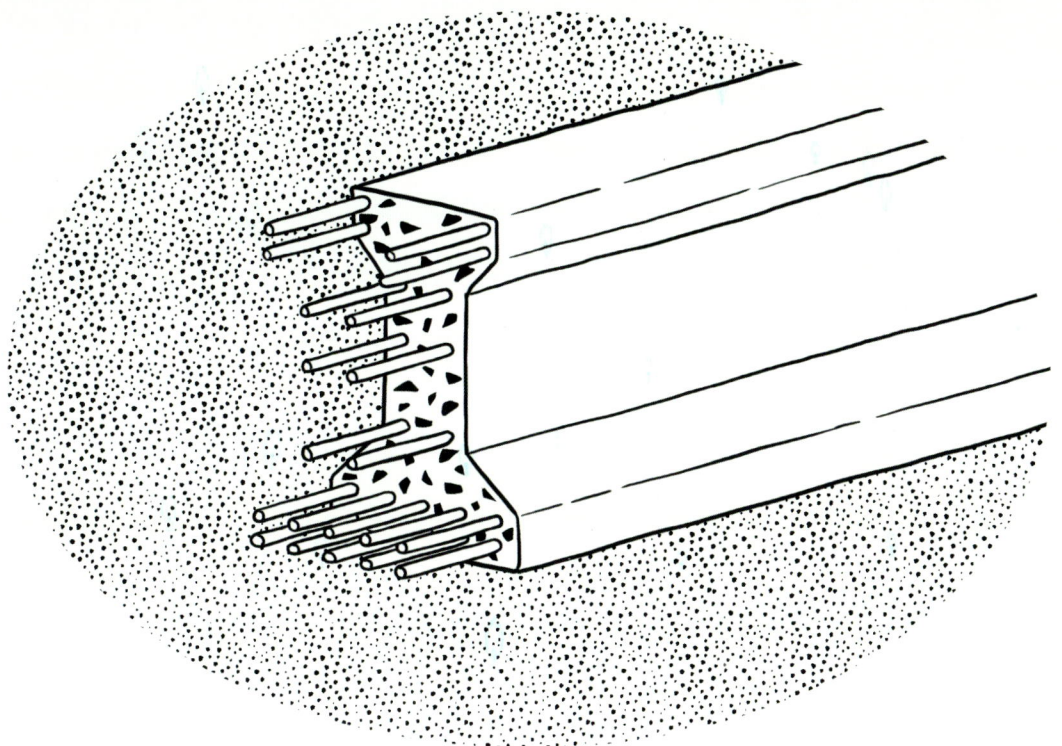

One popular material that makes up a large proportion of bridge beams is concrete. Beams made from concrete last longer and require less maintenance than those made from steel or other materials. But concrete has some weaknesses. Although strong when compressed, concrete quickly fails when subjected to tension forces, those trying to pull it apart. Bridge beams experience both tension and compression forces, so they must be able to resist both simultaneously. That's why structural members made from concrete are reinforced with steel bars. Reinforcement within concrete creates a composite material, with the concrete providing strength against compressive stress while the reinforcement provides strength against tensile stress. For bridge beams, this reinforcement is often *pre-stressed*. The steel bars are stretched and held taut while the wet concrete is cast into a mold. Once the concrete hardens, the tension in the steel compresses it tightly like a rubber band, making the beams stiffer and less prone to cracking. These beams are constructed in factories so they can show up to a job site ready to be lifted into place.

4

BRIDGES AND TUNNELS

Introduction

For all its natural beauty, the Earth often creates difficulties in getting around. In fact, many of the most magnificent of Earth's features are also the most challenging to traverse. Rivers and mountains aren't conducive to roadways, railways, or other corridors along the ground's surface. When the topography is too wet, steep, treacherous, or prone to disaster, the only way forward is up or down. At canyons, valleys, and rivers, our roadways break free from the earth using bridges. And at hills, mountains, and shallow waterways, they bore right through to the other side. Perhaps because these structures solve such a grand but singular problem—creating a path to the other side—bridges and tunnels are among the most celebrated human achievements and full of fascinating engineering details. They are almost always custom-designed for a specific location, conforming to the local topography, geology, and hydrology (not to mention regional architectural preferences and styles). So, each bridge and tunnel is distinct with its own individual character. Because of their size and importance, these structures often reflect that character outward, becoming a symbol of the places they connect.

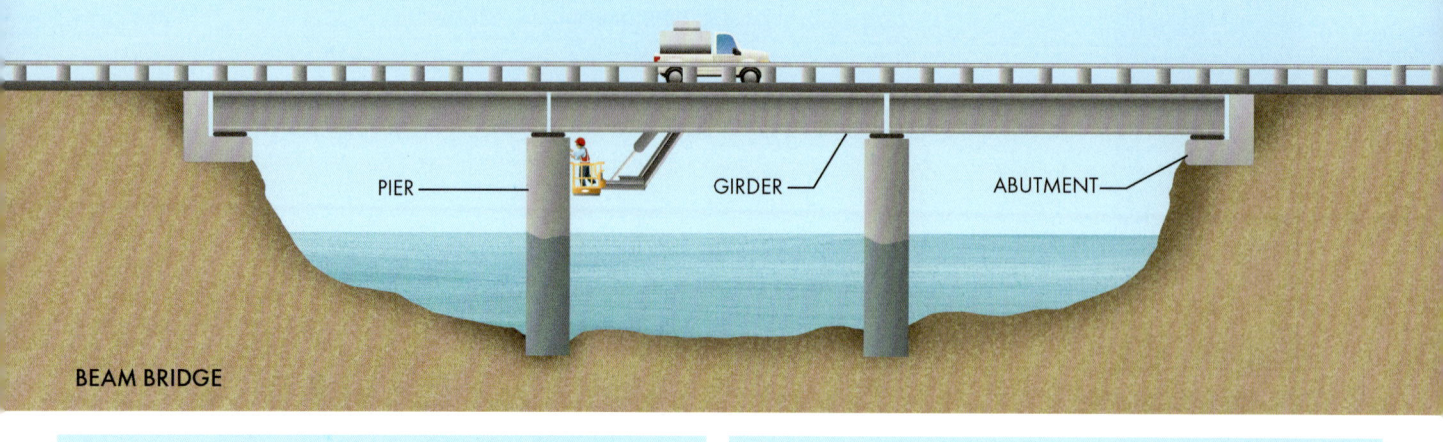

BEAM BRIDGE

PIER — GIRDER — ABUTMENT

DECK — TRUSS

TRUSS BRIDGE

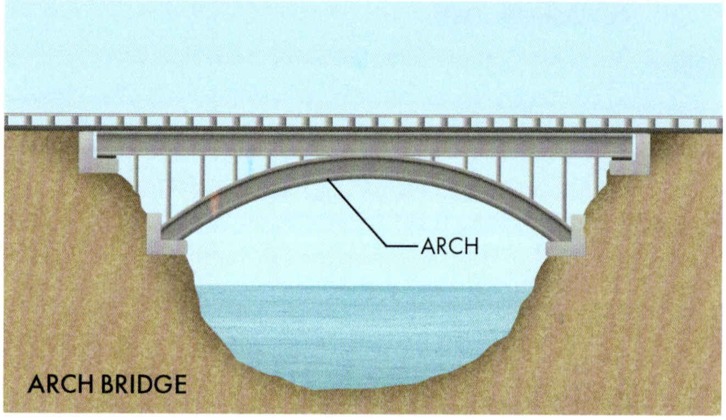

ARCH

ARCH BRIDGE

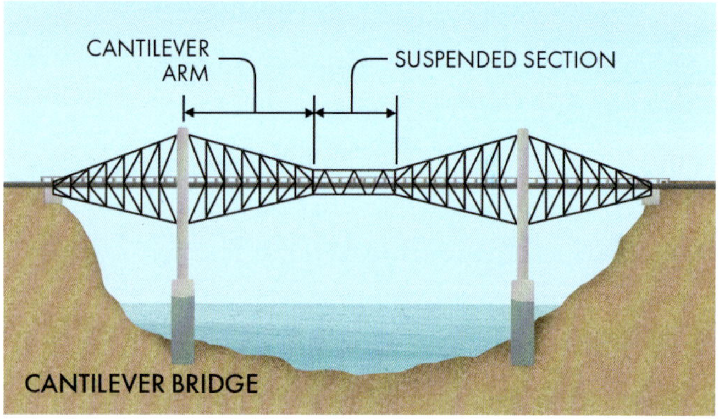

CANTILEVER ARM — SUSPENDED SECTION

CANTILEVER BRIDGE

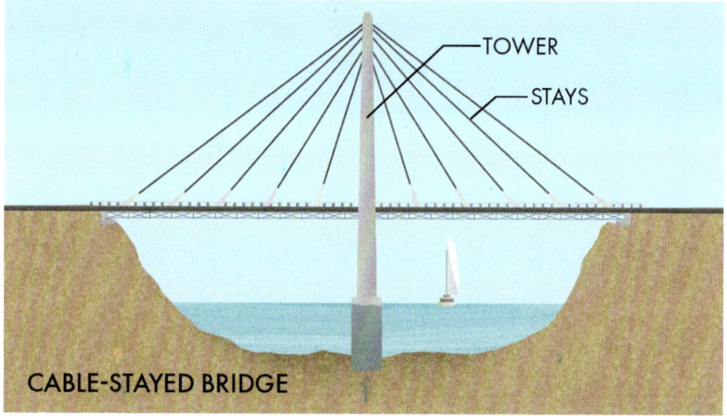

TOWER — STAYS

CABLE-STAYED BRIDGE

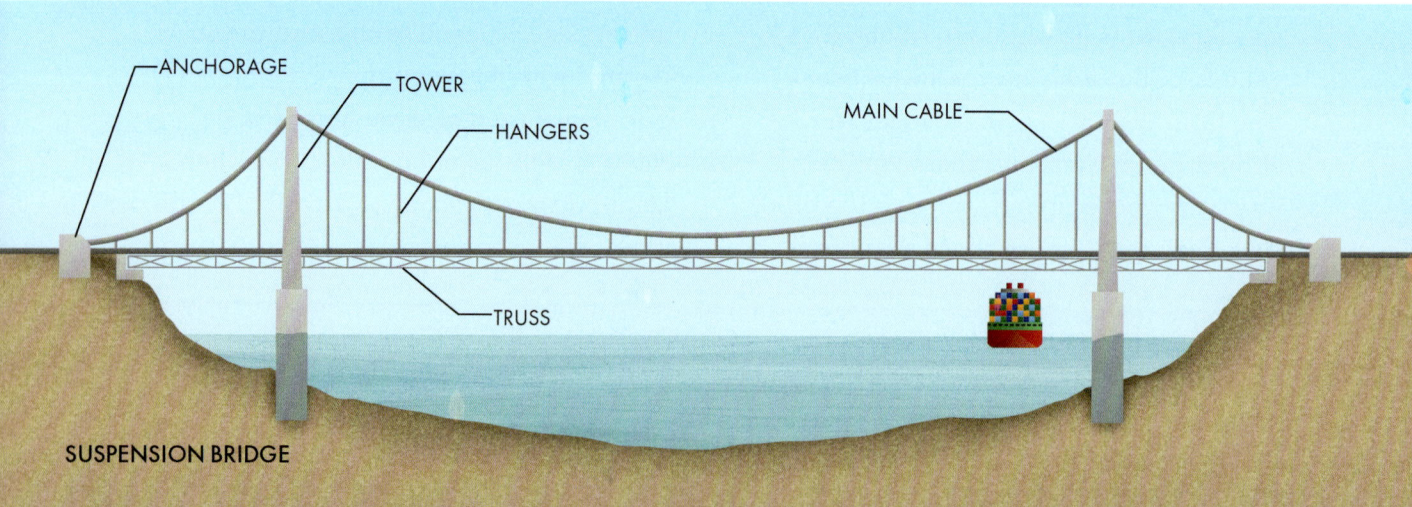

ANCHORAGE — TOWER — HANGERS — MAIN CABLE

TRUSS

SUSPENSION BRIDGE

Types of Bridges

Much of the infrastructure we rely on day to day is not necessarily picturesque. We certainly could build exquisite electrical transmission lines or stunning sanitary sewers, but we rarely want to bear the cost. Bridges are different, though. It seems that humanity decided if we must clutter up the most scenic parts of the landscape with structures, we should at least give them some charm. That's not to say ugly bridges don't exist in the world, but the physical appearance of bridges is often a significant consideration during their design. To an enthusiast of constructed works, many are downright breathtaking. There are so many ways to span a gap, all singular in function but remarkably different in form. No matter how it's accomplished, there is something magical about a structure that can support substantial loads with nothing underneath.

One of the simplest structural crossings is the BEAM BRIDGE. It consists of one or more beams (often called GIRDERS) resting on PIERS or ABUTMENTS below. Beam bridges usually can't span great distances because the girders required would be too large. At a certain distance, the beams become so heavy that they can hardly support their own weight, let alone the roadway and traffic on top. Beams are mainly used for short bridges or in applications that allow for many intermediate piers for support. Most bridges used in highway interchanges are of the beam type. Although beautiful in their own

way, overpasses are usually quite utilitarian. (See Chapter 3 for more details on interchanges.)

One way around the challenge of the structural members' self-weight is to use a TRUSS instead of a girder. A truss is an assembly of smaller elements that creates a rigid and lightweight structure. This weight reduction allows trusses to span greater distances than solid beams. TRUSS BRIDGES can take many forms. The illustration shows a *through truss*, with the road DECK on the bottom level and structural members above the bridge (as opposed to a *deck truss*, which hides the structural members below the road).

Another type of bridge takes advantage of a structural feature that has been around for millennia: the ARCH. Most materials are stronger against forces along their axis than those applied at right angles (called *bending forces*). ARCH BRIDGES use a curved element to transfer the bridge's weight to abutments using compression forces almost exclusively. Many of the oldest bridges used arches because it was the only way to span a gap with materials available at the time (stone and mortar). Even now, with the convenience of modern steel and concrete, arches are a popular choice for bridges. They make efficient use of materials but can be challenging to construct because the arch can't provide its support until it is complete. Temporary supports are required during construction until the arch is connected at its apex from both sides.

When the arch is below the roadway, we call it a *deck arch bridge* (like the example shown). Vertical supports transfer the load of the deck onto the arch. If part of the arch extends above the roadway with the deck suspended below, it's called a *through arch bridge*. Arches can be formed in many ways, including individual steel beams, steel trusses, reinforced concrete, or even stone or brick masonry. One result of compressing an arch is that it creates horizontal forces called *thrusts*. Arch bridges usually need strong abutments at either side to push against that can withstand the extra horizontal loads. Alternatively, a *tied arch bridge* uses a chord to connect both sides of the arch like a bowstring so it can resist the thrust forces. If each end of the arch sits atop a spindly pier, you can be sure that they are tied together.

Another way to increase the span of a beam bridge is to move the supports so that sections of the deck balance on their center instead of being supported at each end. A CANTILEVER BRIDGE uses beams or trusses that project horizontally from their supports, moving most of the weight above the supports rather than in the center of the span. A typical cantilever bridge has four supports, of which the two central piers bear the compression loads from the bridge. The outermost supports resist tension to provide the balancing force for each CANTILEVERED ARM. Cantilever bridges often use large steel trusses, but they can be constructed of concrete as well. Some even include a SUSPENDED SECTION between the two cantilevered arms.

The longest bridges in the world take advantage of steel's ability to withstand incredible tension forces. CABLE-STAYED BRIDGES support the bridge deck from above through cables attached to tall TOWERS. The cables (also called STAYS) form a fan pattern, giving this type of bridge its unique appearance. Depending on the span, cable-stayed bridges can have one central tower or a pair. Their simplicity allows for a wide variety of configurations, giving rise to some dramatic (and often asymmetric) shapes.

Where a cable-stayed bridge attaches the deck directly to each tower, a SUSPENSION BRIDGE instead uses two massive MAIN CABLES to dangle the road deck below with vertical HANGERS. Suspension bridges are iconic structures due to their enormous spans and slender, graceful appearance. A TOWER on either side props up the main cables like a broomstick in a blanket fort. Most of the bridge's weight is transferred into the foundation through these towers. The rest is transferred into the bridge's abutments through immense ANCHORAGES keeping the cables from pulling out of the ground. Because they are so slender and lightweight, most suspension bridges require stiffening with girders or TRUSSES along the deck to reduce movement from wind and traffic loads. These bridges are costly to build and maintain, so they are constructed only when no other structure will suffice. Many consider suspension bridges to be the pinnacle of civil engineering ingenuity.

One final style of bridges is those that move, usually to allow passage for boats and ships. Although not common, there are numerous types of moveable bridges around the world, all of which are unique and customized for a specific location.

One of my favorite activities in a new city is to watch a moving bridge to try to figure out how it works.

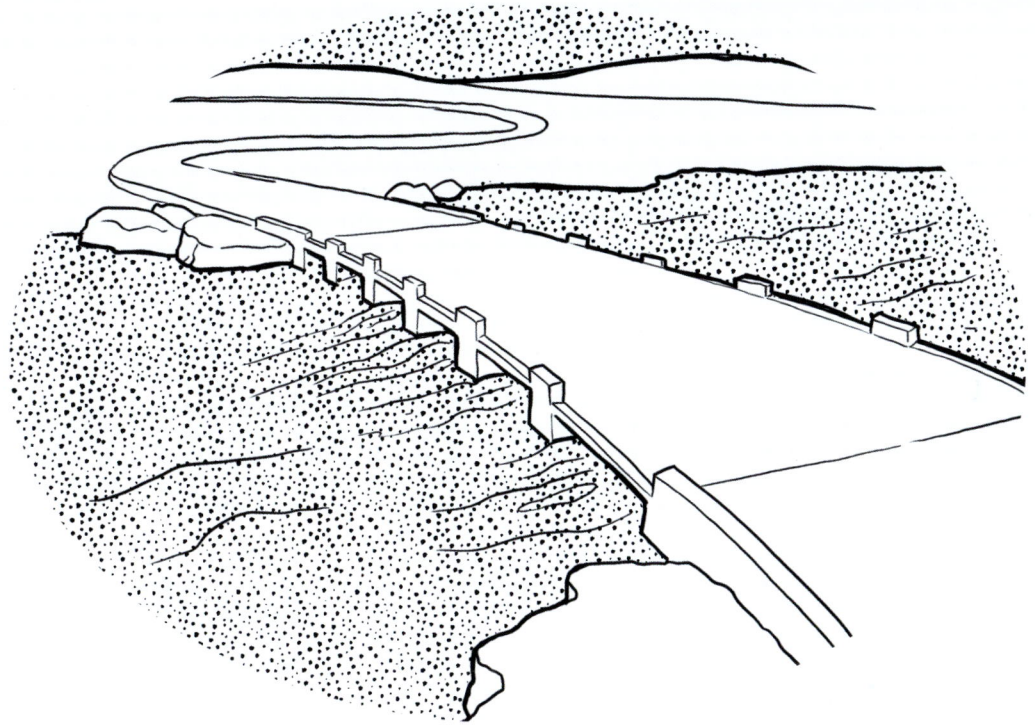

When funding is tight, one option to span a small stream is a *low water crossing*. Unlike bridges built above the typical flood level, low water crossings are designed to be submerged when water levels rise. They are most common in areas prone to flash floods, where runoff in streams rises and falls quickly. Ideally, a crossing would be inaccessible only a few times per year during heavy rainstorms. However, low water crossings have some other disadvantages. For one, these types of bridges can block the passage of fish just like a dam. The other problem with low water crossings has to do with safety. A significant proportion of flood-related fatalities occur when someone tries to drive a car or truck through water overtopping a roadway. Water is heavy. It takes only a small but swift flow to push a vehicle down into a river or creek, which means at least some of the resources saved by avoiding the cost of a bridge often are spent to erect barricades during storms, install automatic flood warning systems at busy crossings, and run advertisement campaigns encouraging motorists never to drive through water overtopping a roadway.

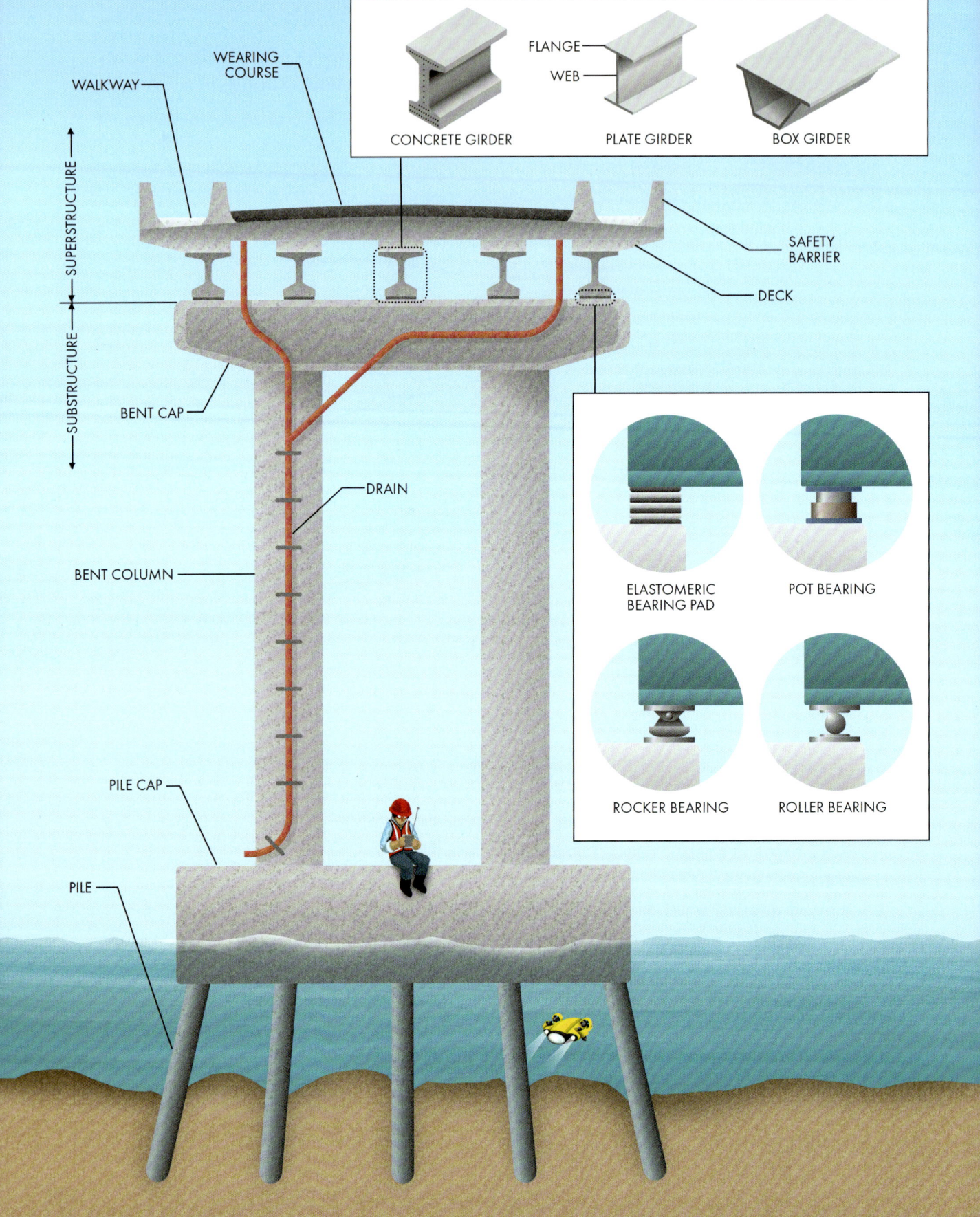

WALKWAY

WEARING COURSE

SUPERSTRUCTURE

SUBSTRUCTURE

BENT CAP

DRAIN

BENT COLUMN

PILE CAP

PILE

FLANGE

WEB

CONCRETE GIRDER

PLATE GIRDER

BOX GIRDER

SAFETY BARRIER

DECK

ELASTOMERIC BEARING PAD

POT BEARING

ROCKER BEARING

ROLLER BEARING

Typical Bridge Section

Although every bridge is different, most share common elements that can be observed from the outside. A cross section through a bridge reveals the individual parts that contribute to its function. Bridges are often divided between the SUPERSTRUCTURE (which carries the traffic loads across each span) and the SUBSTRUCTURE (which transfers the weight of the superstructure into the foundation). Each of the two sections contains fascinating details.

The surface of a bridge upon which vehicles travel is called the DECK. Most often, it consists of a concrete slab set atop the beams. In some cases, the deck is *precast*, meaning the concrete has already been formed and cured before it is lifted into place. Otherwise, the deck is cast in place, using formwork to hold its shape until the concrete hardens. If this method is used during construction, it must be done carefully. Concrete is heavy, after all, and as more and more of it is added to the beams, the structure will begin to flex. To avoid cracks, contractors carefully sequence their work so that most of this movement happens early during placement before the concrete fully hardens.

The deck includes a slope, either from the center (called a *crown*) or from one edge, to ensure that rainwater won't pond on the roadway. A layer of waterproofing and pavement is added to the concrete deck slab to protect it from harsh weather and damage from traffic. This WEARING COURSE also smooths out any unevenness,

providing a more enjoyable ride for motorists. It is meant to be replaced regularly while the slab beneath remains a permanent part of the bridge. A bridge deck also often features SAFETY BARRIERS along the edges to prevent errant vehicles from falling off, DRAINS to direct water away from structural members, and WALKWAYS to accommodate pedestrian traffic.

Most bridges will have some type of beams, or girders, to support the deck, depending on the design. For beam bridges, these are the primary load-bearing elements transferring all the forces to the substructure. For other types of bridges, the beams may only be adding stiffness to the deck or supporting its weight between hangers, cable stays, or the nodes of a truss that do the brunt of the lifting. Girders experience the most significant forces along their upper and lower extremities. Generally, the top of the beam undergoes compression and the bottom experiences tension, so most girders are shaped like a capital "I" to have more material in the FLANGES, with a narrow WEB in the center where forces are not so significant. These girders are usually made from steel PLATES or reinforced CONCRETE. Another popular shape is the BOX GIRDER, which is essentially a closed structural tube. Boxes are often used on bridges that curve because they can withstand twisting better than typical girders.

BEARINGS transfer loads of the superstructure to the substructure; they "bear" the bridge's weight. The girders can't sit directly on the piers or abutments for a straightforward reason: bridges move. The superstructure deforms and vibrates under the moving traffic loads, expands under the glaring sun, and shrinks when it cools down (especially on frigid winter nights). Without isolation from the substructure, these movements would build up stress and potentially cause structural members to fail. Bearings provide this isolation while also reducing wear and tear on the supports by making sure forces are evenly distributed. There are many exciting solutions to these challenges, and if you pay attention, you'll notice a wide variety of bridge bearing styles.

Most modern bridges use an *elastomeric* (in other words, flexible) material to support the weight of the deck and girders while allowing for minor vibrations, rotations, and translations between the piers. Sometimes this ELASTOMERIC BEARING PAD is a standalone component consisting of pure rubber or laminated layers of rubber and steel plates to control bulging. Another option is a POT BEARING, which houses the elastomeric material in a steel cylinder. The pot keeps the rubber from squishing out at its sides, allowing the use of a softer and more pliable material. Pot bearings sometimes include steel plates to accommodate sliding motions and can be designed to restrain or release different motions, depending on the needs of each bridge. Many older bridges used ROLLER BEARINGS or ROCKER BEARINGS to allow for both rotation and horizontal movement

of the superstructure. These types of bearings are mostly being phased out because they are costly to maintain.

The substructure consists of vertical elements that carry the loads from the girders, deck, trusses, cables, and hangers and transfer them to the underlying ground. A substructure can take many different forms, depending on the nature of the soil and rock below the bridge, whether the elements will be subject to powerful scour forces from a river, and the type of bridge being supported. Solid intermediate supports are usually called *piers*. Alternatively, when a support consists of multiple COLUMNS with a CAP, it is called a BENT. At each terminus of the bridge span is an *abutment*. These supports are often larger than piers or bents because they withstand both vertical and horizontal loads from the superstructure. Abutments also serve as the transition between a bridge and the at-grade roadway, so they sometimes act as a retaining wall for the soil below the approaching roadway.

The bridge's foundation is the part of the substructure that transfers the weight of the piers, bents, or abutments into the earth. Some foundations consist of a simple concrete pad called a footing. However, most bridge foundations use PILES, slender steel, or concrete members drilled or driven into the earth. Sometimes piles are battered (in other words, given an angle from vertical) to help resist horizontal forces in addition to vertical ones. Multiple piles are used at each support, and the group is tied together with a PILE CAP upon which the columns sit.

The bearings between the sub- and superstructures provide support while allowing freedom of motion to avoid building up unnecessary stresses, but bridges also need a gap in the road deck to make room for such movement. This gap is called an expansion joint, and it must be at least as wide as the difference between the bridge's length on its hottest and coldest days. The longer the bridge's span, the wider this gap must be. Motorists and their vehicles don't like to drive over large, unsupported spaces. So, bridge decks include miniature bridges that allow cars and trucks to safely pass the expansion joint. These joints usually feature interlocking steel fingers or a compliant rubber material to close the gap for motorists. Listen for the "clomp-clomp" as you pass over it the next time you're driving on a bridge or elevated roadway.

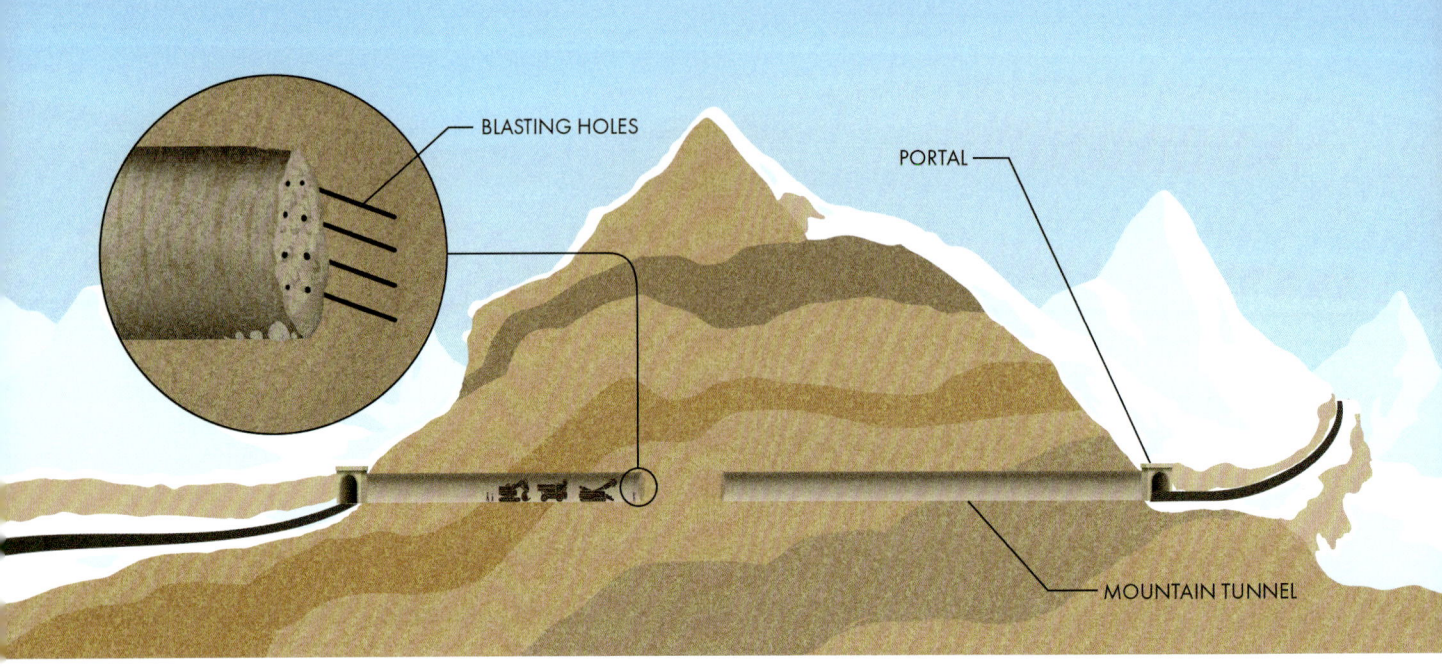

BLASTING HOLES

PORTAL

MOUNTAIN TUNNEL

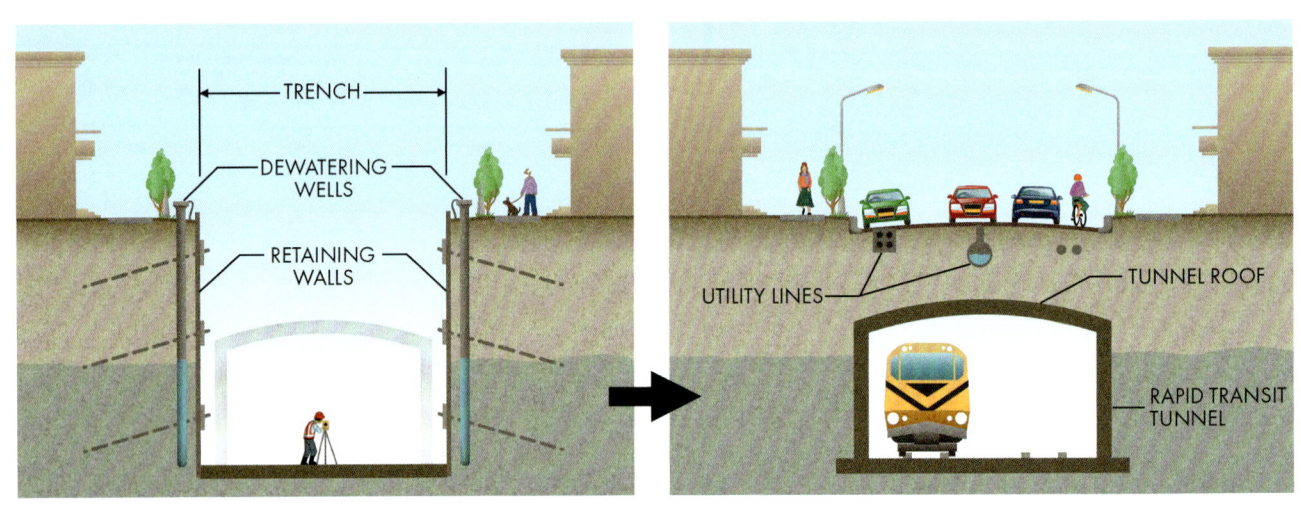

TRENCH

DEWATERING WELLS

RETAINING WALLS

UTILITY LINES

TUNNEL ROOF

RAPID TRANSIT TUNNEL

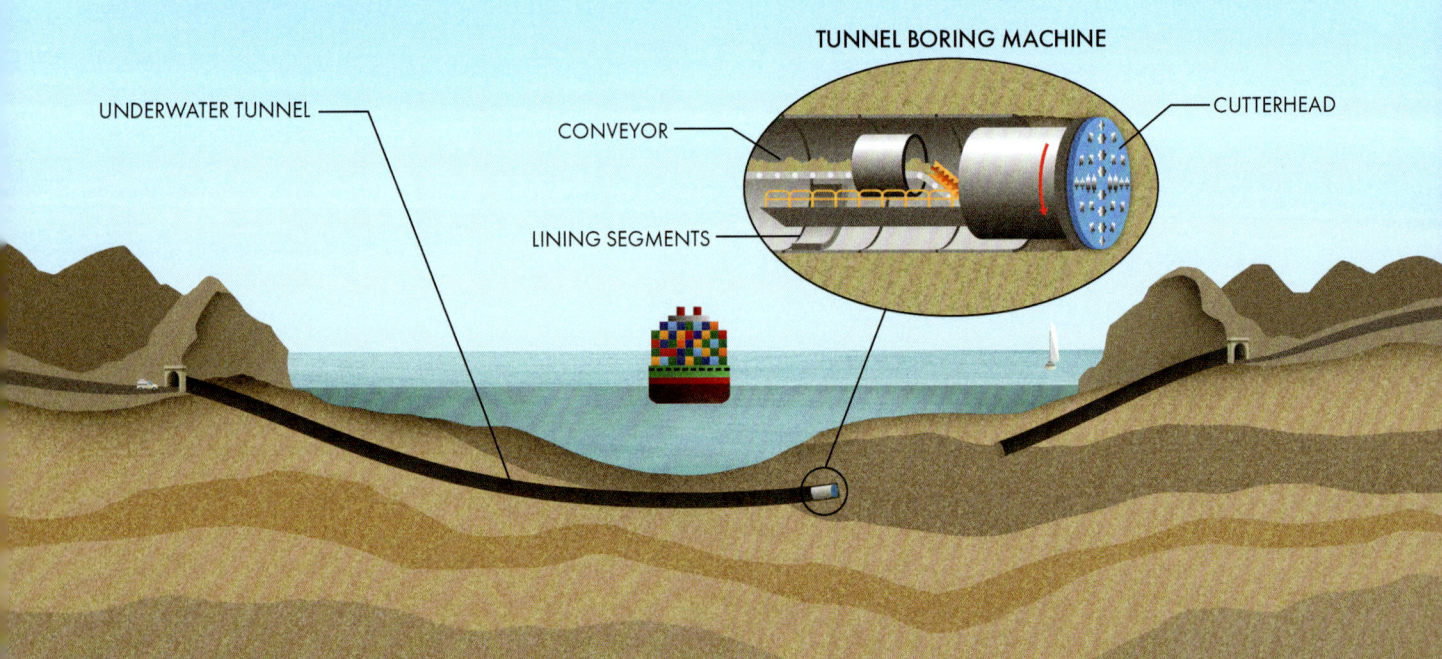

UNDERWATER TUNNEL

TUNNEL BORING MACHINE

CONVEYOR

CUTTERHEAD

LINING SEGMENTS

Overview of Tunnels

The concept of a tunnel is relatively straight-forward: a hollow tube through the earth within which motor vehicles, trains, and even pedestrians can travel. However, tunnels are among the most technically challenging and expensive engineering projects in the world. A few types of infrastructure make use of underground tubes (many of which are discussed within this book), but this chapter focuses on tunnels used for transportation. Although costly and challenging to construct, tunnels enable travel across geographic features that would otherwise be difficult or impossible to traverse. They also open a whole new dimension of travel, maximizing the use of valuable land in dense urban areas. It's a different world below the Earth's surface, both for engineers designing tunnels and the travelers passing through them. But there is something inherently intriguing about navigating through the ground rather than on its surface.

One of the primary jobs of a tunnel is to allow people to cross an obstacle. Tunnels are common in mountainous regions where grades are too steep or treacherous along the surface. Instead of winding up and over the steep topography, sometimes it is more practical to simply go right through it. Some MOUNTAIN TUNNELS stretch only a short distance between POR-TALS (the name for tunnel entrances and exits), but the longest are upward of 30 miles (50 kilometers) long.

Water is another obstacle that can be overcome with a tunnel. Bridges aren't always the simplest way to cross a river or bay, especially in areas with heavy maritime traffic. Where a bridge might encroach on a waterway with its supports, an UNDERWATER TUNNEL allows for unrestricted travel of boats and ships.

Another critical role of tunnels comes in dense urban areas where space on the surface is precious. Rapid transit railways often use subterranean space, allowing them to avoid conflicts with surface roadways and other infrastructure. Because they usually aren't far below the surface, many RAPID TRANSIT TUNNELS are often constructed using the *cut-and-cover* method, starting with a TRENCH. Excavating below the surface of an urban area is a disruptive and challenging ordeal. Existing roadways must be rerouted. UTILITY LINES must be protected or redirected. Nearby buildings may need extra support to avoid settling. RETAINING WALLS are required to hold the trench open while the tunnel can be constructed (more on them in Chapter 3). Finally, groundwater must be continuously managed. If the retaining walls are not watertight, temporary DEWATERING WELLS may be installed to pump it directly from the earth. Another option is *ground freezing*, which uses a refrigeration system and coolant pipes to freeze a layer of water and earth into an impermeable barrier. This temporary wall of ice strengthens the soil and prevents the migration of groundwater into the work area.

Once the trench is excavated, the features of the tunnel itself can be constructed, whether rail or roadway. The ROOF is the final element to be installed. After that, the trench is backfilled, and the infrastructure on the surface can be restored.

Cut-and-cover methods are often used to build underwater tunnels as well. In *immersed tube construction*, prefabricated tunnel sections can be carefully sunk into dredged trenches below the water. Each unit is attached by divers, backfilled with soil to prevent floating, and then pumped dry. In urban areas, cut-and-cover tunnels are usually constructed in short sections because it isn't feasible to open a long stretch of the ground in a city for months or years. Such disruption is avoided by the other method of tunnel construction: *boring*.

Just like cut-and-cover, boring a tunnel follows a few main steps: excavate and remove the soil or rock, install supports to hold back the surrounding earth and water, then complete the tunnel features. The benefit of boring is that it can be performed without disturbing the surface, speeding up construction and making it possible to build in areas that would be otherwise inaccessible (such as along busy streets or below existing buildings). Although historical methods of tunnel construction used a variety of techniques, modern tunnels are bored in two primary ways. First, they can be manually excavated. In rock, the face of the tunnel is advanced by drilling BLASTING HOLES, filling the holes with explosives, and blasting the material apart. In soft soils, crews may use a temporary support called

a *shield* to provide access to the tunnel face. A significant benefit of manually excavating a tunnel is that the design can be adjusted to accommodate changing geology. Additional support is installed only when needed (for example, when the rock is weak or fractured), saving the cost of unnecessary strengthening.

The other option is to use a TUNNEL BORING MACHINE (TBM). These massive pieces of equipment act like giant drills, using a rotating CUTTERHEAD to chew through the rock and soil. TBMs also include CONVEYORS to remove the spoils as they are excavated and equipment to install concrete LINING SEGMENTS that support the tunnel walls and roof. (The next section includes more details on tunnel lining.) Although they are hugely expensive and difficult to transport, these machines can make tunnel construction a rapid and efficient process. They are most often used on long, large-diameter projects or tunnels in very challenging ground conditions.

Tunnel excavation tends to be a slow process, so longer tunnels are sometimes constructed from both sides at the same time. This cuts down on construction time but creates a challenge. How can two crews blindly bore toward each other and meet accurately in the middle? The surveyors who guide tunnel construction crews or a TBM in the right direction do not have access to navigational satellites or surface reference markers. Instead, they often rely on the Earth's magnetic field to establish a bearing. A magnetic compass isn't accurate enough for this purpose because of the interference from iron and steel used in construction. Even a tiny error in direction can compound to significant deviations over a long distance. So, surveyors make use of gyroscopes that can point toward north to a high degree of precision. These instruments make it possible for tunnels to accurately break through in the center of the exit shaft and even for two tunnel construction crews to meet in the middle.

KEEP AN EYE OUT

Driving into a tunnel during the day creates a sharp transition from the bright sunlight outside to the artificial illumination inside the tunnel. Engineers call this the *black hole effect*. It can be a serious safety issue because human eyes adjust to changes in brightness gradually. Drivers can be blinded by the sudden darkness at the entrance of a tunnel and the subsequent brilliance at the exit. Many creative solutions have been employed to solve this luminosity predicament. Some tunnels use shade structures ahead of each portal to provide a smoother transition in illumination. Some use white paint on the walls at the entrance and exit to reflect more of the artificial light into the driver's vision. Most modern tunnels simply use custom lighting to ensure motorists see clearly along the entire length. Pay attention, and you'll notice the intensity of the lights gradually changing from bright to dim and bright again as you pass through.

EXHAUST
AIR DUCT

INTERIOR
WALL

VENTILATION
DAMPERS

EMERGENCY
EXIT

EXIT

EVACUATION
CORRIDOR

SUPPLY AIR
DUCT

DRAINAGE
CHANNEL

LINING RING

MACHINE-BORED TUNNEL

ROOF
LINING

WALL
LINING

JET FAN

FINAL
LINING

SHOTCRETE
LINING

DRAINAGE
CHANNEL

CUT-AND-COVER TUNNEL

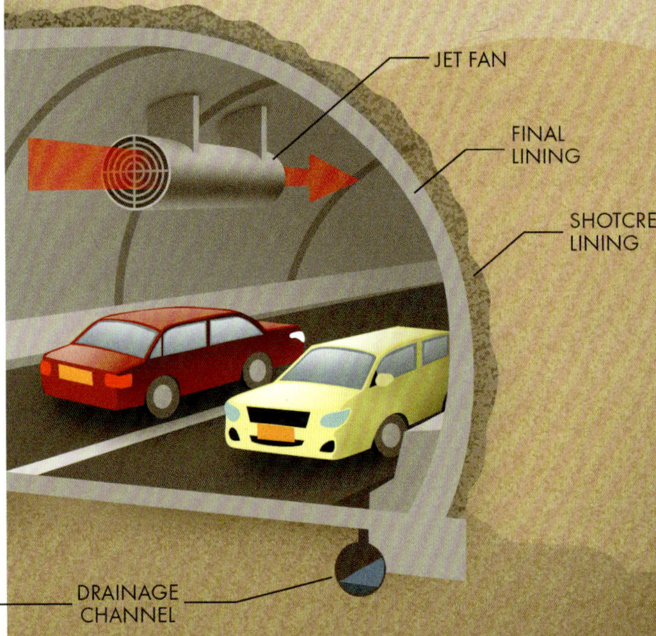

MANUALLY BORED TUNNEL

Tunnel Cross Section

Every tunnel is a unique structure designed for a specific situation. It may not seem like there is much room for variety when it comes to digging passageways through the ground. However, many considerations can affect a tunnel's design, including location, length, depth, geology, traffic volume, and more. Many details make it possible to travel in passageways through the earth in safety and comfort, and they are fun to spot if you know what to look out for.

Like atmospheric pressure created by the weight of air, pressure also exists in the earth's subsurface from the mass of the soil and rock above. This pressure compresses the subterranean material more and more the farther down you go. Building a tunnel through the earth interrupts the flow of these compressive forces. Similar to removing a column from a building, excavating a tunnel takes away the support. Tunnels are often constructed below the groundwater table, making them subject to water pressure as well. But where loads in a building come only from above, earth and water pressure in a tunnel can come from all sides. Most tunnels are installed with a *lining* to resist the pressure from the ground, hold the passageway open against collapse, and minimizing the infiltration of groundwater.

MANUALLY BORED TUNNELS are often lined using concrete sprayed on the walls, called SHOTCRETE, to provide the initial support. This layer helps hold the soil and rock together while stresses redistribute after excavation. A FINAL LINING of steel or concrete is added later. In urban CUT-AND-COVER TUNNELS, the lining usually consists of reinforced concrete cast in place. The falsework and reinforcing steel are erected first, and then concrete is pumped or poured into the forms to harden. Once it has cured, the formwork is removed, and the soil around the tunnel WALLS and ROOF can be backfilled. For MACHINE-BORED TUNNELS, the lining usually consists of concrete RINGS. Each ring is made from precast segments and delivered to the tunnel face, ready to be lifted into place. The segments include a gasket to seal out groundwater and use tapered geometry to lock tightly together when installed.

Most tunnels have an arched or circular cross section because it's the strongest shape against ground pressure. The arch redistributes the forces around the passage just like an arch bridge over a river. However, a tunnel may not look circular to a motorist since many use INTERIOR WALLS to separate traffic from the various support systems and utilities. Although they are often hidden from view, a careful observer can see hints of these systems when traveling through a tunnel.

A critical function of a tunnel's support systems is drainage. There must be a way to manage precipitation that comes in through the portals, groundwater that seeps through the lining, and water used to wash the tunnel walls or for fighting fires. Drainage usually enters a CHANNEL or pipe through slots in the roadway curbs. When possible, a tunnel can be sloped such that water drains from the center toward the portals. However, many tunnels are too deep underground to drain freely. In this case, they are equipped with small reservoirs at low spots called sumps. When the sump fills with water, a switch turns on a pump that delivers the tunnel drainage to a sewer or outfall. The water in a tunnel often picks up pollution as it travels, so it can become quite dirty. Modern tunnels often include ways to treat drainage before discharging it.

One of the most vital safety elements of a tunnel is ventilation. Engines, tires, and brakes emit a range of pollutants that can be confined and concentrated inside a tunnel. Also, vehicles occasionally catch fire. When this happens in a tunnel, the resulting smoke can be particularly hazardous because there aren't many options for egress. Managing the flow of air into and out of a tunnel is quite complicated. Too little ventilation will allow pollution to build up. However, excessive airflow can accelerate the growth of a fire and create turbulence that prevents smoke from rising. Tunnels employ many different ventilation schemes to keep the fresh air flowing.

Many tunnels work like simple pipes with fresh air entering one portal and exhaust air leaving the opposite side. This scheme is known as *longitudinal ventilation*. It is accomplished using JET FANS mounted to the ceiling that force the air inside the tunnel to keep moving. Another option is to blow a jet of air into the tunnel entrance at a shallow angle through an opening called a *Saccardo nozzle*. Longitudinal ventilation

works best on tunnels with a single direction for traffic flow because the air moves along with the vehicles. During a fire, the cars beyond the accident can exit the tunnel along with the airflow carrying away smoke. Vehicles caught upstream of the fire are also upwind, so they aren't exposed to harmful smoke.

Above a certain length, longitudinal ventilation becomes less efficient. It is challenging to create enough pressure to keep the air moving effectively over exceptionally long distances. And, even if the airflow is sufficient, it picks up pollutants as it travels such that its quality at the end of the tunnel is much poorer than at the entrance. In these cases, it makes more sense to use *transverse ventilation* where the air is supplied or exhausted at discrete locations throughout the tunnel's length. Transverse ventilation requires DUCTS to deliver fresh air or remove exhaust from each DAMPER in the tunnel. Two ducts are necessary for a fully transverse system: one for SUPPLY AIR and one for EXHAUST. The newest ventilation systems use zones that can extract smoke from a fire without transporting it along the length of the entire tunnel. Sophisticated control systems can identify accidents and adjust dampers and fans to isolate each zone.

Many tunnels feature EMERGENCY EXITS to ensure motorists can reach a safe place in case of an accident or fire. The well-marked doorways lead either to an adjacent parallel tunnel or a protected EVACUATION CORRIDOR. Ventilation keeps the evacuation routes pressurized so that smoke can't enter even when the doors are open.

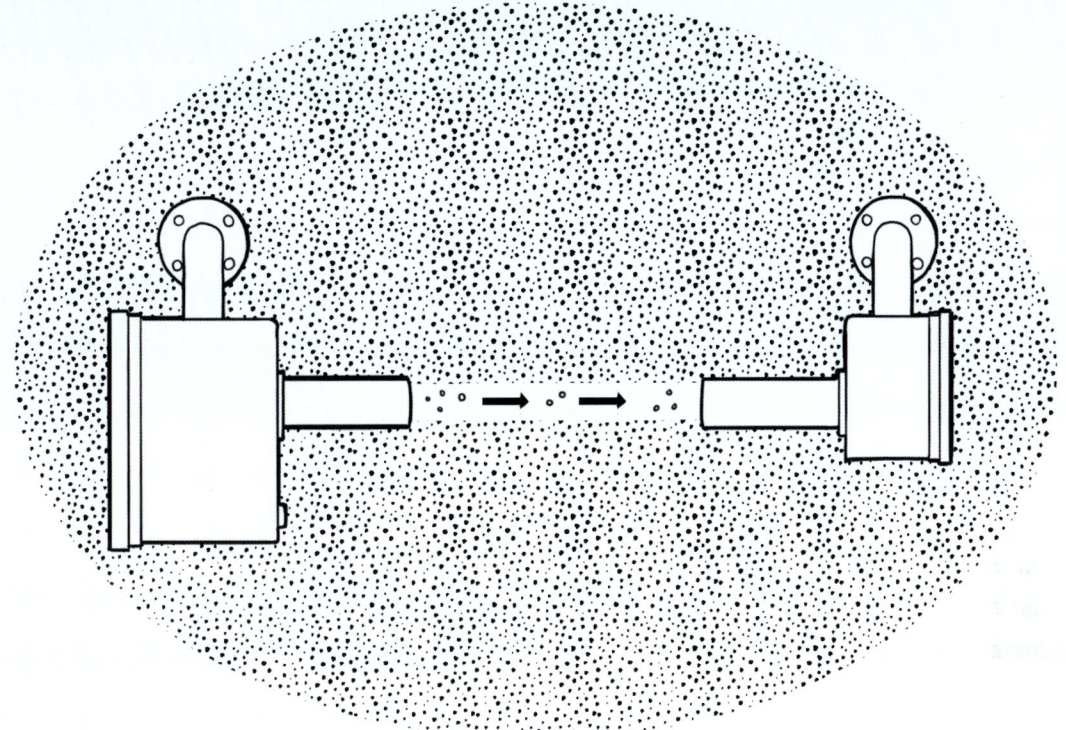

A tunnel ventilation system must be adjustable to ensure that enough fresh air is delivered no matter the volume of traffic or the emergencies that arise. Many designs work like the thermostat in your house, except instead of using temperature, they measure air pollution. When the quality of the atmosphere inside the tunnel starts to decline, the monitoring system increases fan speed or opens dampers to refresh it with outside air. However, measuring pollution takes a lot more ingenuity than measuring temperature. Many air quality sensors in tunnels use light to sense the concentration of dangerous gases. An emitter sends out an intense beam through the tunnel's ambient air. The receiver measures the intensity of the light a short distance away. Many kinds of pollution can reach hazardous levels inside a tunnel, and each one has a fingerprint revealed by the specific wavelengths of light it absorbs. The receiver uses complex algorithms to estimate the concentration of many different gases with high accuracy. This process is called *spectroscopy*, and the monitoring devices that use this principle have a distinct appearance. Look for a pair of enclosures with cylindrical light shields facing each other from a short distance away.

5

RAILWAYS

Introduction

Railroads are one of the earliest forms of overland travel, and they are woven into the history of just about every country in the world. In the United States, railroads fueled enormous expansion and economic growth, perhaps more than any other technology in the 19th century. Today, railroads continue to serve as a vital mode for moving freight and people from place to place.

Railways take advantage of two features to provide rapid and efficient transportation of people and goods. First, the steel wheels on steel rails waste little energy to friction (especially compared to rubber tires on asphalt). Locomotives may look huge, but their engines are almost trivial compared to the enormous weight they move. If your car were so efficient, it could run off a tiny string-trimmer engine.

Second, and more important, railways run along dedicated rights of way, relatively direct and unobstructed paths that are not affected by motor vehicle traffic. These reserved tracks create a level of reliability that is hard to match with other modes of travel.

More than any other type of infrastructure, railways enjoy a contingent of devoted enthusiasts around the world (often self-described as *railfans*). Whether it's due to nostalgia for an earlier era or the simple appeal of seeing large machinery up close, enjoying the details of railroads is a passion for many, and there is much to enjoy. Perhaps more than the trains themselves, the paths they travel along are full of noteworthy particulars ripe for observation and appreciation.

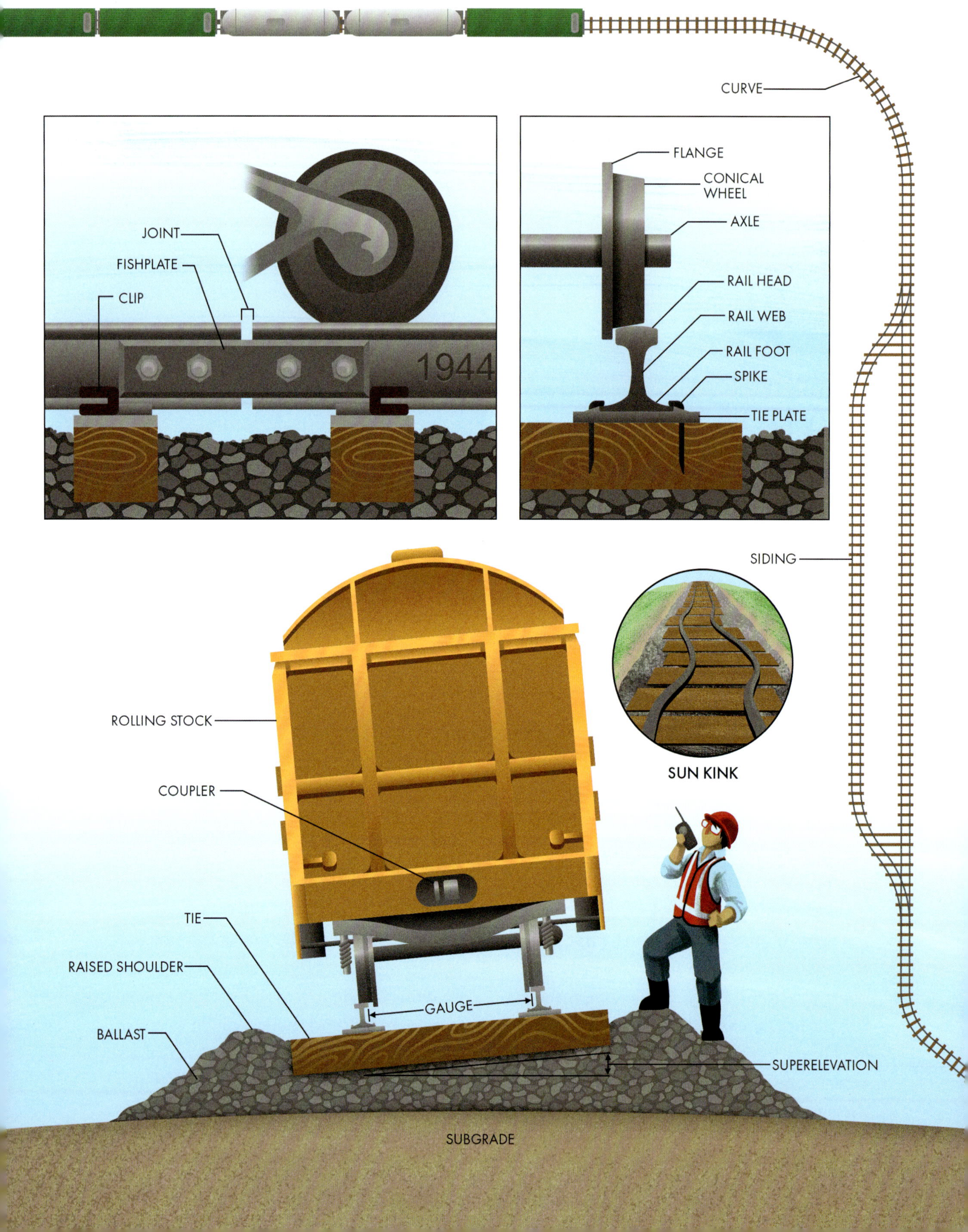

CURVE

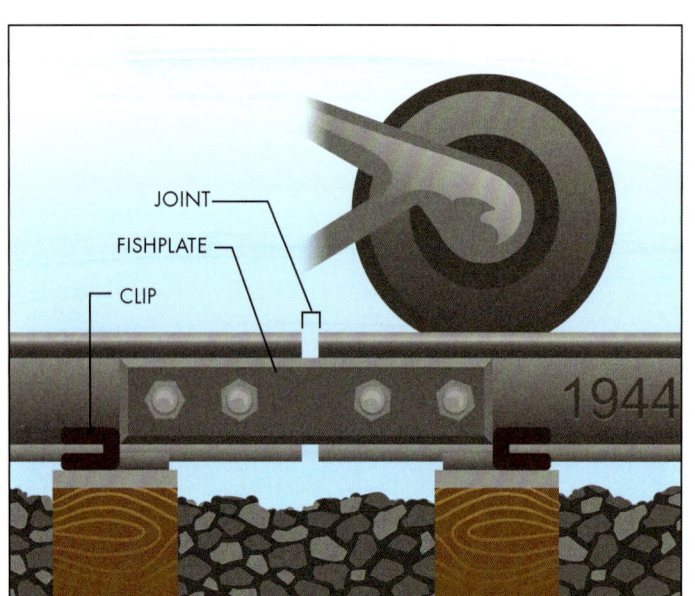

JOINT

FISHPLATE

CLIP

1944

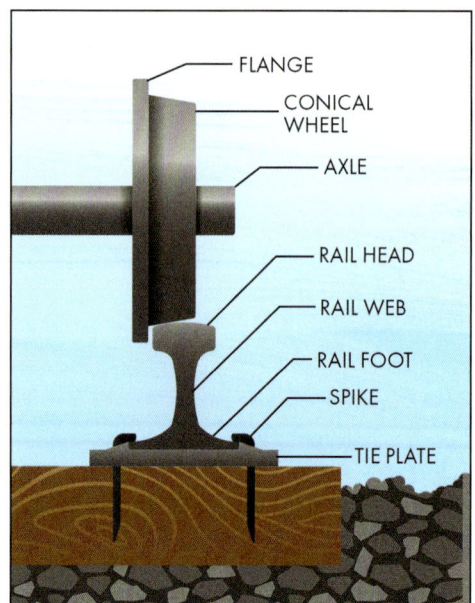

FLANGE

CONICAL WHEEL

AXLE

RAIL HEAD

RAIL WEB

RAIL FOOT

SPIKE

TIE PLATE

SIDING

SUN KINK

ROLLING STOCK

COUPLER

TIE

RAISED SHOULDER

BALLAST

GAUGE

SUPERELEVATION

SUBGRADE

Railroad Tracks

Railroad tracks consist of all the elements required to deliver train traffic quickly and smoothly to its destination. The most distinct aspect of a railway is the rail itself, which supports the tremendous weight of trains and cargo. Rails are made from high-quality steel to withstand these incredible stresses. Look closely, and you can often note markings on the RAIL WEB that list the year of manufacture and other details about how each rail was made. Rails can be of different sizes and shapes, but they mostly follow a similar form: an I-shape with a bulbous HEAD upon which wheels ride and a flat FOOT attached to the ties.

The force required to move a train forward is transferred into the rail through friction with the driven wheels of the locomotive. Incredibly, the contact patch between each wheel and the rail is only the size of a small coin. That means an average freight train sits on an area of steel roughly the size of this book.

Historically, lengths of rail were bolted together using FISHPLATES. The JOINT between each section creates the iconic click-clack sound as the wheels of the train pass over the small gap. These small but frequent discontinuities generate wear and tear on the railway vehicles (called ROLLING STOCK) and are uncomfortable for passengers. Most modern tracks use welded rails to create continuous stretches of smooth-running track with no joints.

One challenge of eliminating these gaps is thermal movement. Steel contracts with low temperatures and expands with heat. Where many structures provide freedom to move using expansion joints, tracks with continuously welded rail restrain this thermal movement. On cold days, the rails experience tensile stress as they try to contract. On warm days, they undergo compressive stress as they try to expand from their restraints. At some point in between, called the *neutral temperature*, the rail is free of thermal stresses. If the ambient temperature deviates too far from the neutral temperature, the stresses can exceed the track's strength. On scorching days, railways can buckle (also called SUN KINK), creating the danger of a derailment. Rails are often warmed or stretched before installation to mitigate the possibility of buckling. This technique raises the neutral temperature of the rail so that hot days don't overload it with thermal stress.

There are many ways to attach the rail to the horizontal TIES (also called sleepers). Historically, a large steel SPIKE with an offset head was hammered in place to secure each side of the rail. These spikes are still used on some railroads in the United States. More modern railways use one of many types of heavy-duty CLIPS. In North America, ties are usually made from wood because of its abundance, but they can be made from concrete as well. Ties have two essential jobs: bear the weight of loads from the train traffic above and keep the two rails running with the correct space between (called the track GAUGE). Wooden ties often include a TIE PLATE to distribute the concentrated force of the rail.

Maintaining an accurate gauge is critical because of the way trains stay on the tracks. You might think a train would struggle to navigate around a bend with a solid AXLE because the outside wheel would need to turn farther than the inside wheel. Cars use a *differential* between the drive wheels so they can rotate independently around curves. Rolling stock gets around this problem by using CONICAL WHEELS. When the train goes around a bend, each axle shifts, so the outside wheel rides on a larger radius and the inner wheel on a smaller radius. This compensates for the difference in travel distance between the inside and outside of the curve. The wheel's FLANGE is only a safety feature to keep it on a track that is damaged or out of alignment. During regular operation, it shouldn't touch the rail at all.

Railroad ties do not sit directly on the ground below the track, called the SUB-GRADE. The soils are rarely strong enough to bear the immense weight of train traffic. Instead, an embankment of loose rock called BALLAST is used to spread the load evenly to the underlying soil. Ballast is often made from crushed stone because the angular features help it interlock into a solid foundation. Not only does it distribute the vertical forces from the track, but ballast also provides horizontal support to each tie, helping resist buckling due to thermal stress and shifting from horizontal train forces around bends. Many embankments have RAISED SHOULDERS to provide extra resistance against lateral forces in each tie. The open spaces within the stone ballast allow water to flow freely through (instead of ponding up along the sides).

The geometry of railways is a critical component in their design. Railroads can use a much narrower right of way than a highway because they don't need large, clear zones on each side of the travel lanes. However, trains require much gentler CURVES and grades than can be navigated by motor vehicles. COUPLERS between cars can't handle sharp bends. Also, the centripetal force around curves can create undue stress on passengers and cargo. One solution to this problem is similar to a feature used on highways: raise the outside rail so that the train leans into curves. This tilting, also called SUPERELEVATION or *cant*, reduces the horizontal force felt by the train.

As for vertical alignment, trains do not have enough traction on steel rails to brake effectively on steep slopes. Significant uphill gradients also cause trains to slow down, reducing the capacity of the railway. Next time you're driving parallel to a railway, watch the tracks as you travel. Although the road will often follow the natural ground closely, the tracks will maintain a much more consistent elevation with only gradual changes in slope.

The number of tracks is another essential consideration in railway design. A single track is less expensive to build and maintain than two but has some disadvantages. Most important, trains traveling in opposite directions must have a way to pass each other. A SIDING (or *passing loop*) is a short section of parallel track that allows trains to pass. The capacity of a single-track railroad depends on the number of these sidings.

Careful scheduling can maximize the use of the single track, but two or more tracks dramatically increase the railway's capacity and reliability.

KEEP AN EYE OUT

Although modern railroads mostly use continuously welded rails, occasional breaks between long sections are still required. This is especially true at bridges or *viaduct* structures that expand and contract at different rates than the tracks above. At these intermittent locations where thermal movement in the rails is unrestrained, the joints must have room for significant deviations in length. Using a butt joint between rail sections would create a major discontinuity for passengers and rolling stock. Instead, *expansion joints* on rails (sometimes called *breather switches*) use diagonal tapers. This oblique joint allows train wheels to transition smoothly from one section of rail to another while still leaving enough room for thermal movement.

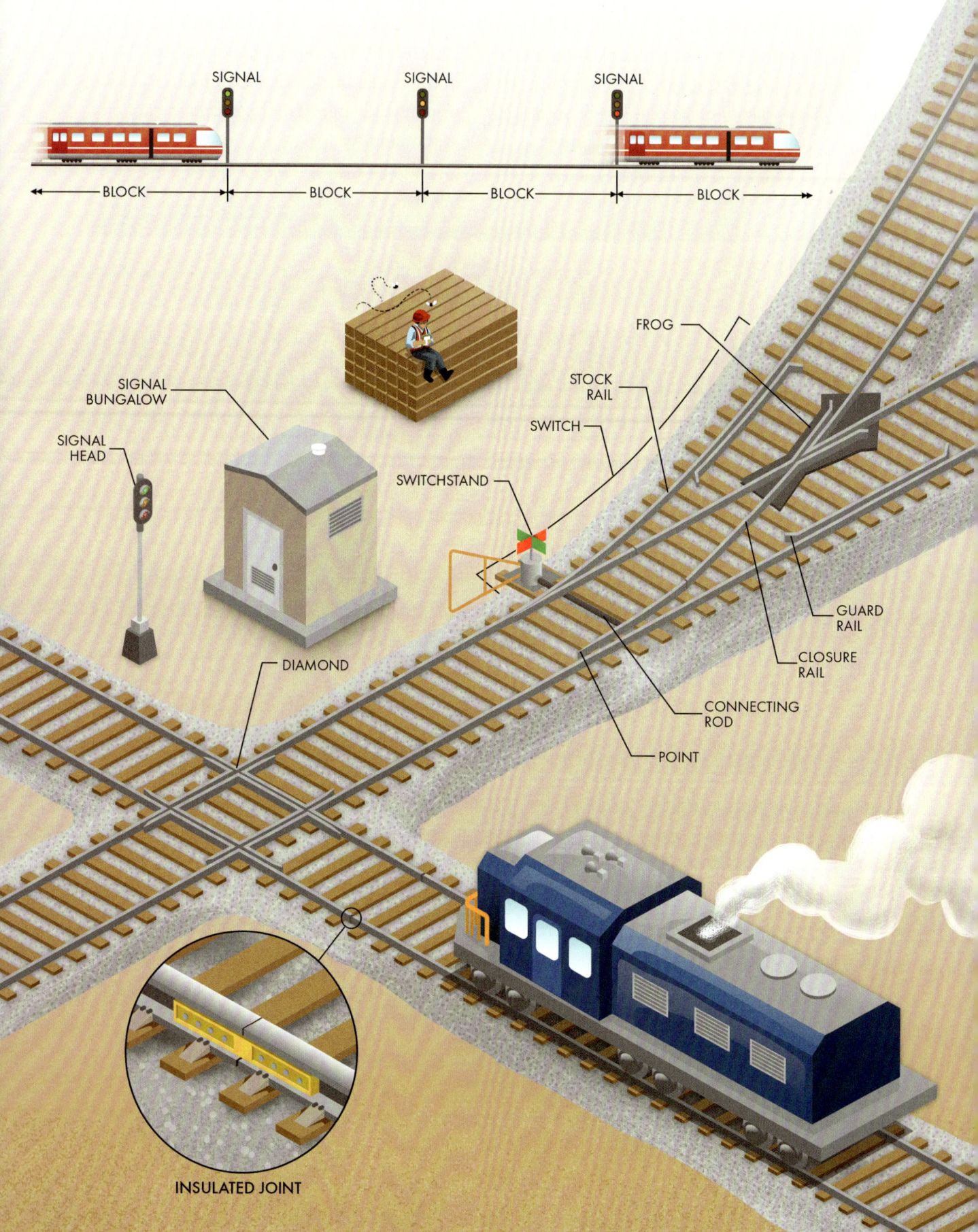

SIGNAL   SIGNAL   SIGNAL

BLOCK   BLOCK   BLOCK   BLOCK

FROG

STOCK RAIL

SWITCH

SIGNAL BUNGALOW

SWITCHSTAND

SIGNAL HEAD

GUARD RAIL

CLOSURE RAIL

DIAMOND

CONNECTING ROD

POINT

INSULATED JOINT

Switches and Signals

Confining trains to their tracks might seem to eliminate the challenge of managing traffic flow. After all, there aren't many opportunities for decision-making when you can move in only one of two directions. However, efficient use of a railroad requires that many trains must share the same tracks. Allowing trains to interact and navigate around each other requires some ingenuity precisely because railways are so constrained to a single dimension.

One significant challenge to managing railway traffic is the considerable distance required to stop a fully loaded train. Unlike motor vehicles whose drivers see and respond to hazards in real time, a train can take more than a mile to reach a complete stop. If a train operator can see an obstruction on the tracks while traveling at full speed, it is already too late. Trains sharing a railway need to maintain enough distance between each other to stop as required without the potential for collision, and they need to keep this distance without relying on the vision of the train crews.

Over the years, many solutions to multi-train traffic management have been used. The earliest method was simply to establish a timetable of when and where each train should be at any time of the day. The apparent limitation of this system is the possibility of a train breaking down or encountering a problem that prevents it from following the schedule. In the best case, a breakdown would delay all the other trains on the line, and in the worst case, it

could lead to a crash. Most modern railway traffic control schemes are instead based on a block system. The tracks are subdivided into segments (called BLOCKS), and trains are prevented from entering any specific block until it is free from obstructions. For non-signalized railways, traffic can be managed through *warrants*. A dispatcher provides a standardized authorization to the crew for specific train movements on the main tracks. However, most heavily trafficked lines use SIGNALS as the primary means of controlling traffic between blocks.

Just like a traffic signal on a roadway (as discussed in Chapter 3), railway signals tell the train operator when it's safe to proceed. In fact, many railway signals use combinations of lights to provide further information about the routes and speed limits ahead. Even in North America, many railways use different standards, so interpreting the meaning of their signals can take some work. The simplest signals are those between blocks, which usually have a single SIGNAL HEAD with three lights—green, yellow, and red—similar to those used at roadway intersections. A green light means the following blocks are clear and the train can continue at full speed. Yellow indicates that the next block is clear, but the one after is obstructed, and the next signal will indicate stop. A red light means the next block is occupied, and the train cannot proceed.

Some signals are controlled by a dispatcher, but many operate automatically using *track circuits*. In the most basic configurations, a low-voltage electric current is introduced into the rails at one end of the block. At the other end, a relay measures the current to control the nearby signals. When a train enters a block, the wheels and axles create a conductive path between the rails, shorting the circuit and de-energizing the relay. INSULATING JOINTS are installed between each track block to ensure that adjacent signals aren't inadvertently triggered. A nonconductive material is used to attach two sections of rail while keeping them electrically isolated. Modern track circuits can even provide information about the location and velocity of each train. The relays, electronics, and batteries used to control the signals are usually hidden inside enclosures called SIGNAL BUNGALOWS.

Beyond block signals, multiple signal heads and light combinations carrying various meanings add even more complexity. The busiest companies use centralized traffic control offices that operate like air traffic controllers to coordinate the schedules and routing so that conflicts are avoided. Modern traffic systems provide warnings and information inside the cab of each train, reducing the possibility of human error. In addition, the most sophisticated signal systems allow trains to communicate their positions to each other so the blocks can travel along with the train, rather than being static stretches of track on a map.

Another vital element of railway traffic management is movement between tracks.

Trains often need to pass one another, divert to destinations off the main line, and swap cars and compartments in rail yards. Without a way to transition between railways, trains would be forever stuck on a single track, making these tasks impossible. SWITCHES (also known as *turnouts*) provide the means for trains to change tracks. The most basic type of switch uses two flexible tapered rails called POINTS. The train wheels are guided in one of two directions, depending on which of the two points is contacting the nonmoving STOCK RAIL. A CONNECTING ROD below the track attaches the points to a mechanism that selects the train direction. Sometimes a SWITCHSTAND with a lever that a rail employee must throw manually controls the switch. Alternatively, a dispatcher may control the switch remotely using an electromechanical *switch machine*.

Once past the points, the train wheels navigate onto one of the two tracks. However, before reaching the main track, the left wheel must cross the right rail of the opposite track or vice versa. These crossings require a gap in the rail through which the wheel flanges can pass, a task accomplished with a FROG. The crossing wheel is handed off from one of the CLOSURE RAILS to the frog as the flange passes through the gap. Adjacent to the frog are GUARD RAILS. These run parallel to the main rails to keep wheels in alignment and protect against derailment. You can also see guard rails used on sharp curves and along bridges.

When two tracks cross each other without a connection between, a DIAMOND is installed. These crossings consist of four frogs to allow each wheel to traverse both rails of the intersecting tracks. Both switches and diamonds receive a significant amount of wear and tear from regular train traffic. When they cross over gaps and joints, wheels generate massive impact forces that can damage the rolling stock and the railway itself. As a result, switches and crossings receive extra attention from inspectors to reduce the likelihood of a failure that could lead to a derailment.

Although rail transportation remains a vital way to move goods and people worldwide, the heyday of rail construction has passed. Over time, consolidation of the rail transport industry and the increased efficiency of other modes of travel have led to the closing of railways in many countries. Luckily, with their gentle slopes, connections to city centers, and passage through beautiful countryside, unused rail corridors are ideal for an alternate use: walking and biking. *Rail trails* convert abandoned lines into long multiuse paths, and they can be found around the world. The longest rail trails extend for hundreds of miles with connections to neighborhoods, parks, stores, restaurants, and even campsites.

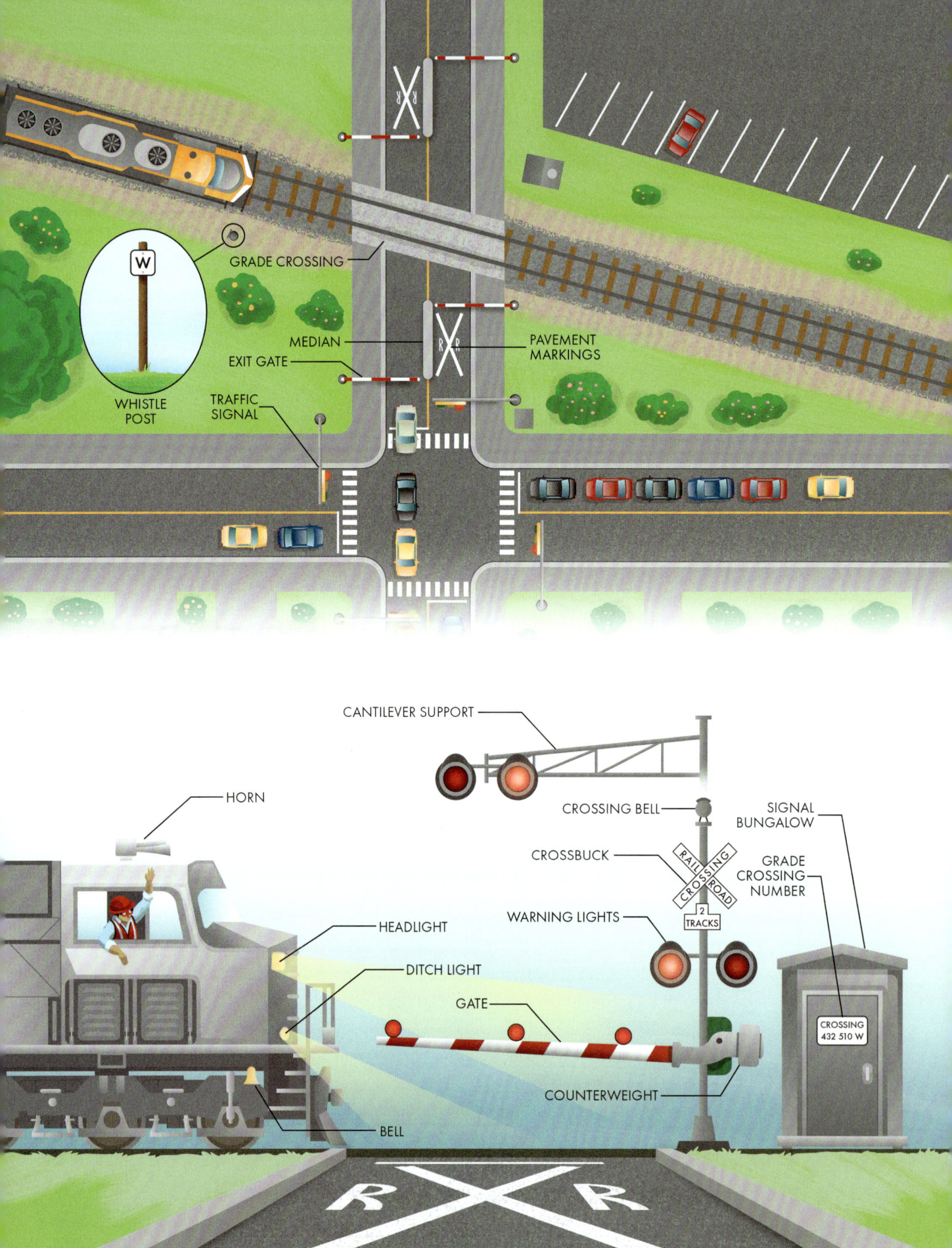

GRADE CROSSING

MEDIAN

EXIT GATE

WHISTLE POST

TRAFFIC SIGNAL

PAVEMENT MARKINGS

CANTILEVER SUPPORT

HORN

CROSSING BELL

SIGNAL BUNGALOW

CROSSBUCK

RAIL ROAD CROSSING

2 TRACKS

GRADE CROSSING NUMBER

HEADLIGHT

WARNING LIGHTS

DITCH LIGHT

GATE

COUNTERWEIGHT

BELL

CROSSING 432 510 W

Grade Crossings

Railroads stretch vast distances across areas completely unpopulated by people, but between those empty stretches are the city centers they connect. The closer a railroad gets to a populated place, the more conflicts with other infrastructure it encounters. Most important, railroads are an impediment to the flow of pedestrian and vehicular traffic. Some roads and railways use bridges to cross each other without interruption, but many intersect at the same level. These GRADE CROSSINGS are where the average person is most likely to encounter a railway. Trains moving at full speed cannot stop within the sight distance of an operator, and they can't swerve to avoid a hazard either. For those reasons, they always have the right of way at crossings. Pedestrians and motor vehicles must stop and wait for trains to pass, so grade crossings include many safety features to decrease the potential for dangerous collisions.

In many countries, grade crossings are assigned an identifier, called the GRADE CROSSING NUMBER, to simplify accident and malfunction reporting. Modern railway companies (and their regulators) are dedicated to public safety and respond quickly to reports of problems. The safety features at grade crossings are generally divided into two categories: passive and active. *Passive warning devices* are those that do not change when a train is approaching. They include a stop or yield sign and a CROSSBUCK, the international symbol for a railway crossing that consists of two slats in an X formation. When more than one rail is present, a supplementary plate states the number of tracks at the intersection. A crossbuck is often included as a PAVEMENT MARKING to make sure drivers know the tracks are coming up. Many low-traffic grade crossings use only passive safety features. It is the driver's responsibility to heed these warnings, look out for trains, and proceed only when it is safe to do so.

Active warning devices provide a visual or audible notice that a train is approaching. They are usually triggered by a track circuit of the same kind used in automatic block signaling (described in the previous section). Like railway signals, the relays, electronics, and batteries that control the automatic warning devices at grade crossings are hidden inside enclosures, often called SIGNAL BUNGALOWS. When a train approaches the intersection, a pair of red WARNING LIGHTS begins to flash, letting motorists know they need to stop. If the roadway has multiple lanes, the crossing may include a second pair of warning lights mounted overhead on a CANTILEVER SUPPORT. Mechanical or electronic CROSSING BELLS also provide an audible warning at the crossing for pedestrians or bicyclists who may not see the flashing lights.

In addition to lights and bells, many grade crossings include GATES that drop across the oncoming lanes when a train crosses the roadway. The gates are equipped with reflective tape and lights to make them more conspicuous, even at night. Many intersections include a central MEDIAN to discourage drivers from going around the gate. At the highest-risk crossings, EXIT GATES are often installed for the same reason. They operate on a delay to avoid trapping a vehicle on the tracks. Most grade crossing gates are designed to provide a visual warning, but they aren't strong enough to hold back an errant vehicle. At crossings of high-speed trains, a more robust *barrier gate* may be installed.

One challenge with grade crossings happens in urban areas where signalized intersections are present near the railway. Red lights form a *queue* of vehicles that can back up across the tracks. You should never proceed across a railway until you know it's clear on the other side. Still, drivers queueing at a traffic signal often misjudge the available space and find themselves inadvertently stopped right on top of the tracks. TRAFFIC SIGNALS at busy intersections near grade crossings are usually coordinated with automatic warning devices. When a train is approaching, the signal goes green to clear the queue blocking the tracks.

A key consideration in the design of grade crossings is the *warning time* between the activation of devices and the train's arrival at the intersection. Engineers need to provide enough time for vehicles to clear the tracks or stop, but not so much that impatient motorists assume the devices are malfunctioning and try to bypass the gates. People are naturally distrustful of automatic equipment, and that

wariness is only reinforced when signals take too long to operate or interrupt a journey for no good reason. Engineers consider the volume and types of traffic, the proximity of signalized intersections, the number of tracks, and many other factors to strike a careful balance. The most sophisticated track circuits can estimate the speed of a train to make sure that warning times aren't too long and can even cancel the warning if a train stops before it reaches the crossing.

Automatic warning devices are designed to operate on the failsafe principle. When a malfunction or loss of power occurs, the device reverts to the safest condition (which is to assume a train is approaching). If power is lost, most devices have batteries to power the flashing lights and bells. COUNTERWEIGHTS are carefully adjusted so that the gates will automatically fall when the electricity isn't present to hold them up. Failsafe operation ensures that motor vehicles won't inadvertently cross the tracks if there's a problem with the warning devices.

In addition to the crossing warning devices, locomotives provide their own warnings, including BELLS, bright HEADLIGHTS, and smaller flashing DITCH LIGHTS. Most noticeably, they sound a blaring HORN ahead of each grade crossing. The standard pattern is two long blasts, one short blast, and one final long blast. This sequence is either prolonged or repeated until the train reaches the crossing. If you look closely, you can sometimes see a WHISTLE POST beside the tracks: a short sign placed ahead of a grade crossing to notify the train operator when to begin sounding the horn. In the United States, they usually consist of a small white marker with a capital W.

With so many types of warnings, it might seem that people would notice whether a train was coming before crossing railroad tracks, but hundreds of fatal collisions between trains and motor vehicles happen at grade crossings every year around the world. If you're driving and see the crossbuck, make sure to stop, listen, and look both ways before crossing the tracks.

KEEP AN EYE OUT

One aspect of railroads that is difficult to escape is the noise, especially at grade crossings where each passing train sounds its deafening horn. Excessive noise from trains can be harmful to human health by increasing stress, disrupting sleep, and even causing long-term hearing loss. Trains often pass through densely populated areas where horns can be particularly disruptive. To mitigate this nuisance, many governments have created *quiet zones*, stretches of track where trains do not sound their horn ahead of grade crossings. Extra safety measures are usually installed to make up for the loss of this important audible warning, including signs reminding motorists to look out for trains. Of course, horns must still be used to warn animals, vehicles, or people on the tracks, but quiet zones otherwise make it much more peaceful to live or work adjacent to a railway.

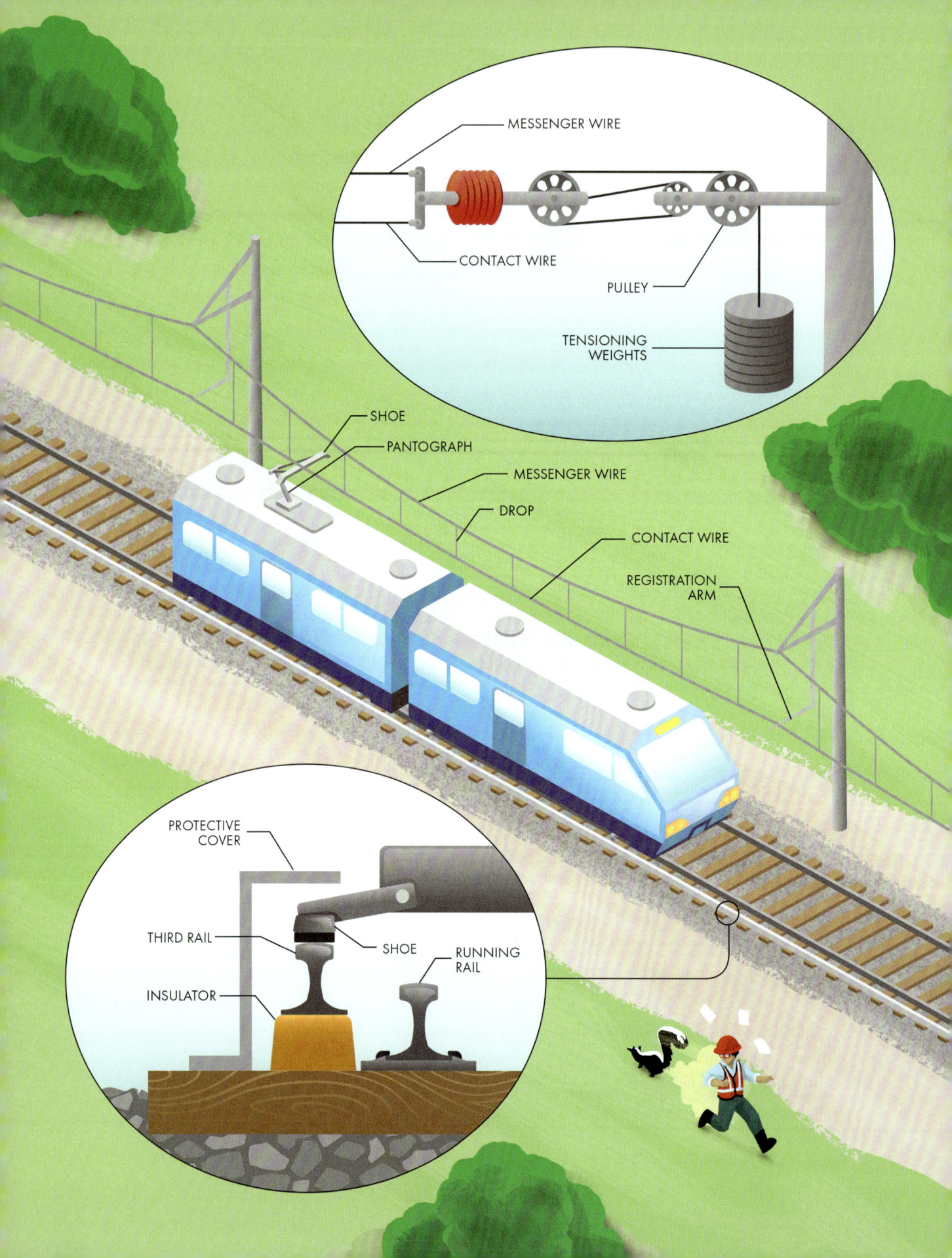

MESSENGER WIRE

CONTACT WIRE

PULLEY

TENSIONING WEIGHTS

SHOE

PANTOGRAPH

MESSENGER WIRE

DROP

CONTACT WIRE

REGISTRATION ARM

PROTECTIVE COVER

THIRD RAIL

SHOE

RUNNING RAIL

INSULATOR

Electrified Railways

Nearly all modern trains run on electric power. Even the large diesel engines in freight locomotives connect to an electric generator that powers *traction motors* to pull the train. Electric motors eliminate the need for massive and complex transmission systems that would otherwise be required to drive the wheels directly from an engine. With the relative simplicity of delivering electricity across long distances, it is natural to wonder why an onboard engine is necessary at all. Indeed, many railways are electrified, meaning they provide electric power for propulsion directly to the train.

Electrifying a railway has many advantages. First, trains don't have to carry the weight of large engines and the enormous volume of fuel they require. They are generally faster and more efficient than their diesel counterparts. Removing the engine also removes its exhaust, improving air quality. This feature is particularly important for trains that run through tunnels or subway systems where engine fumes could concentrate to dangerous levels. Nearly all rapid transit systems use electric railways. Finally, electric trains are capable of regenerating electricity upon braking. Instead of converting the kinetic energy into wasted heat with brakes, the electric motors can act as generators, transforming it back into electricity that other trains on the railway can use. On rapid transit, where trains decelerate quickly, *regenerative energy* comes in short bursts, reducing its usefulness to other trains. However, in areas with many hills, it can be a boon. In an ideal situation, much of the energy a train uses to climb a large hill can be returned to the system as it descends to be used by other trains.

There are numerous electric railway standards around the world, many of which have not changed for more than 100 years. Many systems use direct current because it's easy to change the speed of a DC motor with simple equipment in the cab. However, low-voltage direct current cannot travel far in conductors without significant losses, so most DC railways require regularly spaced substations to convert grid power to direct current along the length of the track. Alternating current can be delivered at a higher voltage and stepped down inside a train. However, it is more dangerous and requires extra equipment onboard the locomotive to convert AC for traction motors.

The infrastructure required to deliver power to moving trains can be quite elaborate, and its cost is the key reason that longer and lower-volume railways are rarely electrified. There are two primary ways to provide electric power to a train: a third rail or an overhead line. THIRD RAIL systems use an energized conductor that runs along the track parallel to the main rails. The energized rail sits atop INSULATORS to keep it isolated from the ground. Trains are equipped with SHOES that slide along the third rail to collect traction power. It's a simple and effective system, but it does create a shock hazard for people or animals near the railway. Strict control of the right of way is required for safety, including fences and warning signs. Many third rails are equipped with PROTECTIVE COVERS to minimize the chance of injury to railway personnel and to keep rain, snow, and ice off the surface.

The other option for delivering electricity to trains is overhead. Overhead lines are safer, and thus, most high-voltage systems are installed above the tracks. In this setup, a current collector sits atop the train. A few different devices can perform this task, but most modern trains employ a PANTOGRAPH. They use spring-loaded arms to maintain contact between a replaceable graphite shoe and the overhead conductor. It's a simple concept but complex in practice. Take a look at standard overhead electric or utility lines, and you'll notice the challenge right away. They sag in the middle of the span. Maintaining contact with such large deviations in height between each support at high speeds would be impossible, so overhead railway power systems use a pair of lines to ensure reliable transfer of electricity to the train. The top wire, called the MESSENGER WIRE, is only for support. The curved shape it takes between poles is called a *catenary*, so that name is often used to describe the whole system. From the messenger wire, vertical supports called DROPS connect the CONTACT WIRE below, and that is the line on which the pantograph rides.

The two-wire system allows the contact wire to be held at a consistent height along the tracks, making it possible for a pantograph to slide along it at rapid speeds. Both wires are energized to carry the traction current, and they are often kept under tension with WEIGHTS suspended on PULLEYS on either side of the lines. This tension takes up slack to reduce the sag of the lines as they expand and contract from temperature changes. The tension also increases the speed of waves that travel along the wires. It makes the vibrations smaller and higher-frequency (just like a guitar string) to minimize bouncing, which can create electric arcs each time the contact wire and pantograph separate. The contact wire is held in a horizontal zig-zag pattern with REGISTRATION ARMS so that the pantograph's shoe wears evenly across its width.

An electric circuit requires a loop, so electrified railways need a second conductor to complete the connection. In most electrified railways, return current travels in the steel RUNNING RAILS on which the wheels roll. With a good connection to the earth, the voltage on the rails will stay low enough to avoid presenting a danger to people and animals. However, return currents create several engineering challenges. For one, the rails are where signal circuits typically travel. If the rails are carrying return current, the small track circuit signals get overwhelmed. Electrified railways often use AC track circuits to control the signals. Relays used to detect trains can be designed with filters to pick up specific frequencies and ignore the traction current in the rail.

Another major issue with using rails in contact with the ground as a return path is stray current. The flow of electricity can take unintended deviations into nearby pipelines, tunnel linings, utility ducts, and other metallic structures. These stray currents can lead to rapid corrosion if not mitigated. Some railways use a fourth rail or additional overhead conductor to provide a return path that is less likely to stray into nearby metallic objects.

KEEP AN EYE OUT

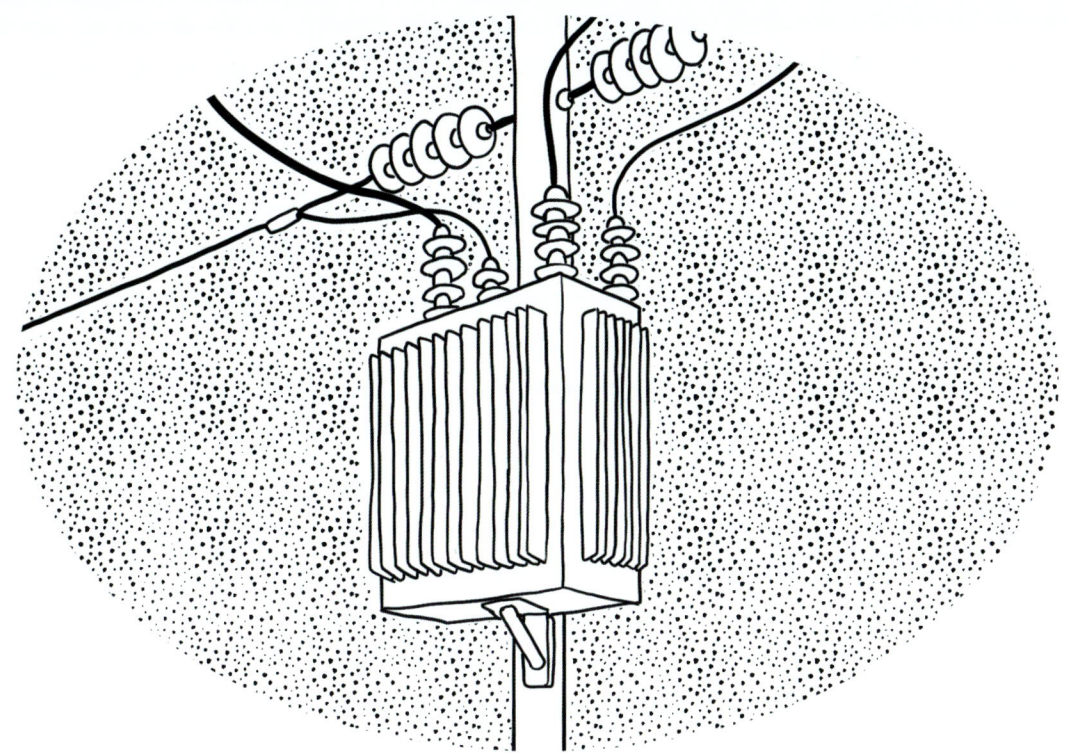

In addition to stray current, AC systems with overhead conductors create a large loop when the return current travels through the rails. These loops generate electromagnetic fields that can induce noise and voltage on communication lines running parallel to the tracks, including those carrying signal information. You never want a red light inadvertently turning green from electrical noise! So, booster transformers are often installed at regular intervals to force the return current into overhead lines, reducing the size of the loops and canceling much of the potential interference.

6

DAMS, LEVEES, AND COASTAL STRUCTURES

Introduction

Like the air we breathe, it's easy to take for granted that our lives practically revolve around water. Not only is water a physiological necessity, but it also serves as a source of power, a means of transportation for goods and passengers, and an excellent place for recreational activities. It serves as a habitat for a multitude of aquatic plants and animals as well. On the other hand, water can be destructive, creating floods that damage property and jeopardize public safety, and eroding riverbanks and shorelines. With its absolute necessity and ever-present threat, it's no surprise that much of our infrastructure is devoted to controlling and managing water.

Many of the world's largest and most complex projects were designed and constructed to either protect against or take advantage of the Earth's immense resources of water. We've built enormous dams to create reservoirs that store freshwater, vast networks of waterways for maritime navigation, and gigantic flood control and coastal protection features around the world. Many of these facilities even attract enough attention and public interest to warrant their own tourism centers that provide a safe vantage for observation and opportunities to learn about their history and technical details. The next time you pass by a major dam, port, lock, or levee, stop by the visitor center, take a tour, and get the T-shirt!

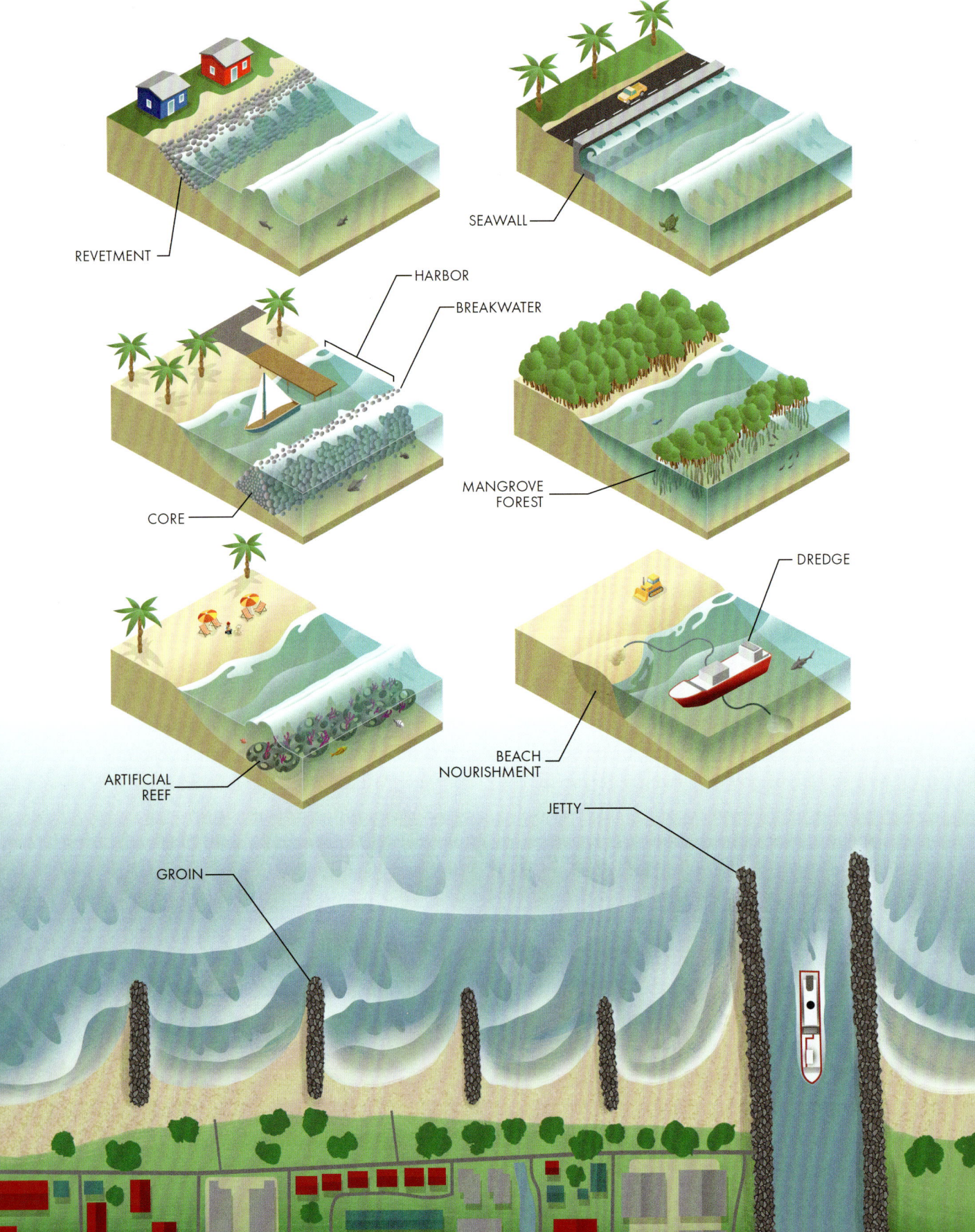

REVETMENT

SEAWALL

HARBOR

BREAKWATER

CORE

MANGROVE FOREST

ARTIFICIAL REEF

DREDGE

BEACH NOURISHMENT

JETTY

GROIN

Shore Protection Structures

Coastlines may look static and unmoving on a map, but they are some of the most dynamic places in the world. Shores are subject to a vast array of natural disruptive forces, including wind, waves, tides, ocean currents, and storms. Humans also impact shorelines by dredging channels, building waterways, developing structures along the coast, and trapping sediment in upland reservoirs before it can reach the shore. It's no wonder that our coastlines shift and transform over time. The soil and rock that make up the coastline are in constant flux, endlessly stolen from one place and deposited somewhere else.

The seashore is essential to humankind, and not just because of the pretty sunsets. Many of our largest cities sit along the coast because of the opportunity provided by shipping and fishing. In addition, beaches support local economies, providing millions of jobs and billions of dollars of economic activity around the world through tourism. Shoreline erosion is a constant menace to our infrastructure, developed areas, and navigational waterways, threatening structures along the shore and the livelihood of vast portions of coastal populations. Much of coastal engineering focuses on ways we can protect the shoreline and combat the disruptive forces that cause it to change and disappear over time.

One of the most basic coastal structures is a REVETMENT, a simple layer of hard armor atop a natural slope. Revetments usually feature large stones or concrete blocks that can withstand the constant force of crashing waves and tidal currents. Using blocks or stones can also absorb the energy of each wave, reducing the distance it travels up the slope. Similar to a revetment, a SEAWALL is a vertical structure parallel to the shore, protecting upland areas against erosion. Seawalls are usually constructed using reinforced concrete. Many seawalls feature a shape called a *recurve* to redirect wave energy back toward the sea, reducing the likelihood of water crashing over the top. Seawalls are usually constructed to an elevation above the normal high tide to protect against flooding and storm surges. They typically separate the developed areas they protect above from the sandy beaches below.

BREAKWATERS are another type of parallel structure used to protect areas of the shore from waves. Unlike revetments and seawalls, they aren't connected to the shore. Instead, breakwaters are constructed offshore to dissipate wave energy and create areas of tranquil water for ships and structures along the coast, called HARBORS. Breakwaters can be made from many materials, but they are most commonly mounds of rock rubble. Often the CORE of the breakwater uses smaller rocks to reduce the flow of wave energy through the structure, while the outside layer consists of larger stones that can better withstand the waves.

Another protective structure, called a GROIN, protrudes into the sea to combat *longshore drift*, the process of sediment movement parallel to the coast. Like breakwaters, groins are usually made using mounds of rock or rubble. Over time, a groin will trap sand suspended in ocean currents to create a beach (a process called *accretion*). If properly sized, a groin can also protect the area on the downdrift side by reducing the speed and power of ocean currents along the shore. However, an oversized groin will rob currents of all their sediment, leaving none to replenish the beaches beyond and thus accelerating erosion along unprotected shores. After one groin is built, additional groins are commonly needed to protect the downdrift area, eventually leading to saw-toothed beaches extending for great distances.

Similar to groins, JETTIES are structures constructed perpendicular to the coast. They are often built in pairs to protect the inlet to a navigation channel by extending its mouth into the sea. Not only do they block the passage of sediment into the channel, but they also confine the flow of seawater within during tide changes, speeding it up to flush sediment from the bottom and minimize its accumulation.

These hard armoring structures often provide a long-term solution to erosion, but they can also create unintended consequences. For just one example, smooth concrete seawalls reflect waves rather than absorb them, potentially worsening erosion farther down the shore. These structures can also affect the quality of habitat in the sea, creating environmental challenges. When possible, coastal engineers look toward "softer" solutions to erosion. One

of those techniques involves planting or maintaining trees and shrubs that can grow in the tidal zones along coasts. These are called MANGROVE FORESTS, and their dense networks of roots absorb wave energy and protect the soil along the coast.

Another soft solution to coastal erosion is to create ARTIFICIAL REEFS that provide habitat for fish, corals, and other marine life. Many materials have been used to construct artificial reefs, including rocks, concrete, shipwrecks, and even submerged subway cars. These reefs offer surfaces where marine organisms can attach or hide, with a secondary benefit of dissipating wave energy offshore, serving as submerged breakwaters.

Yet another soft solution is to reverse the process of erosion by replacing the material that has been lost, a technique commonly known as BEACH NOURISHMENT. Beaches are not only essential recreation areas and economic drivers, but they also serve as buffers between development and the sea. They dissipate the energy from storms and waves before it can reach developed areas, but sand can be moved downdrift or pulled out into deeper waters in the process. Replenishing lost sand protects coastal structures and creates spaces for recreation. Nourishment is often accomplished by borrowing sediments from the seafloor with a DREDGE and pumping them back to the shore as a *slurry* of water and sand in a pipeline. The slurry is discharged ashore into a large basin to allow the water to drain and sand to settle out, after which it can be spread along the beach using earth-moving equipment. Beach nourishment has environmental impacts, and it's not a permanent solution, but it is a popular tool for addressing coastal erosion.

Finally, sometimes the cheapest option to protect shoreline development from damage is for it not to exist in the first place. This strategy, often called *retreat*, involves purchasing and condemning property or relocating buildings and infrastructure farther from the shore. In some cases, the best engineering is to let nature do what it does best: allowing the coastline to be vibrant and dynamic, which is what draws humans to it in the first place.

KEEP AN EYE OUT

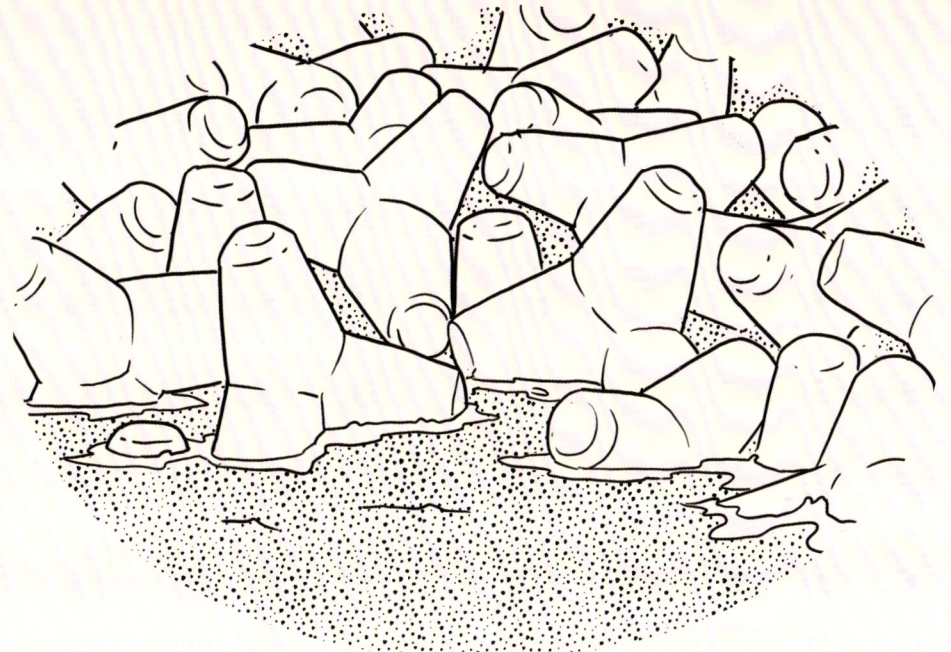

Boulders are a cost-effective way to armor the coastline against the destructive power of seawater, wind, and waves. However, not every part of the shore has a nearby quarry that can provide rocks in such quantity as required for coastal structures. Another option for creating revetments and breakwaters is to use cast concrete blocks, often known as *armor units*. These unique structures are formed in geometric shapes, allowing them to entangle and interlock to resist powerful hydrodynamic forces. A wide variety of concrete armor unit shapes exist. These blocks are often easier to transport and place than unwieldy boulders because they are consistent in size, shape, and weight. They can also be manufactured closer to a project site, reducing transportation costs (especially in areas with no rock quarry nearby).

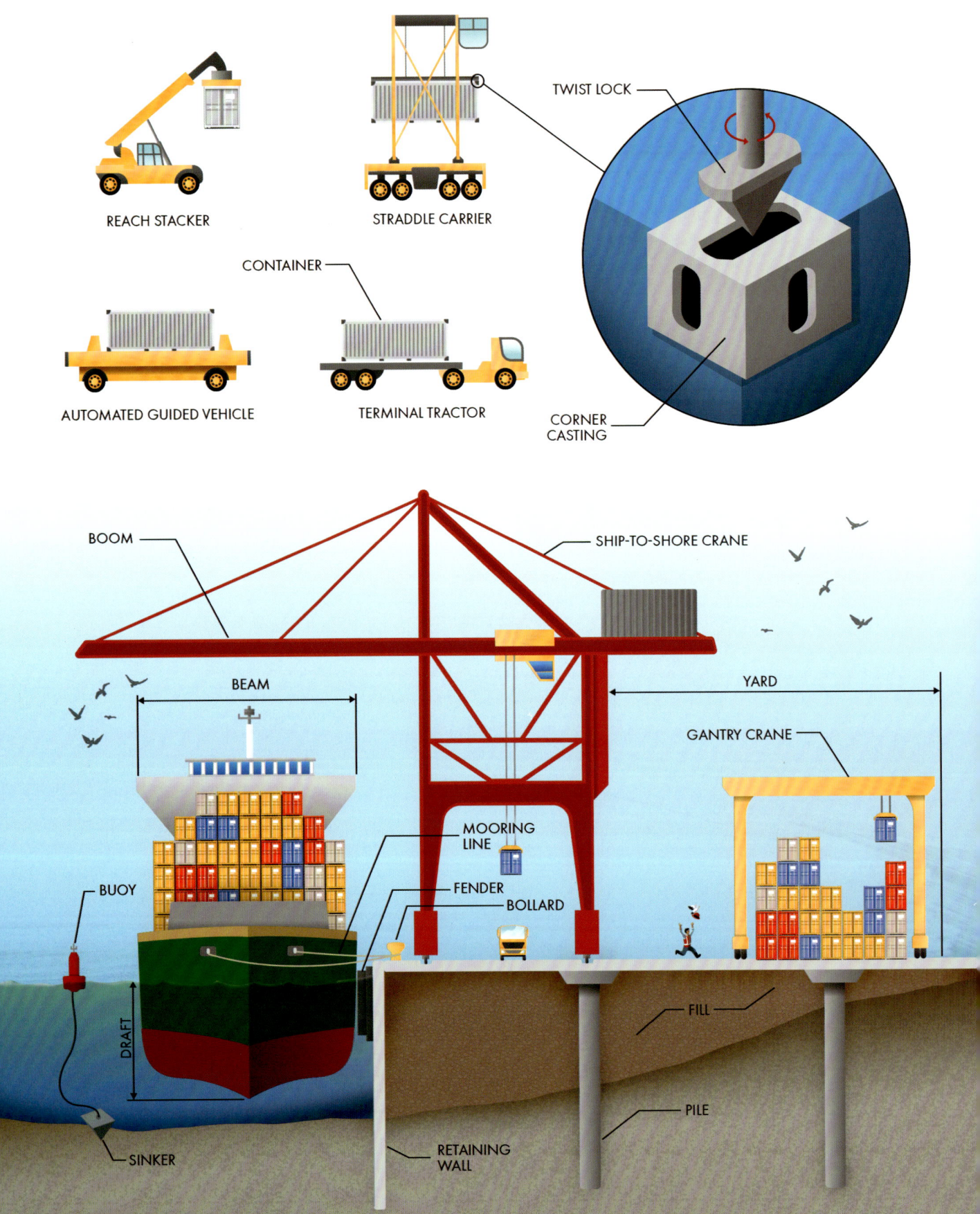

REACH STACKER

STRADDLE CARRIER

TWIST LOCK

CORNER CASTING

CONTAINER

AUTOMATED GUIDED VEHICLE

TERMINAL TRACTOR

BOOM

SHIP-TO-SHORE CRANE

BEAM

YARD

GANTRY CRANE

BUOY

MOORING LINE

FENDER

BOLLARD

DRAFT

FILL

PILE

SINKER

RETAINING WALL

Ports

Maritime transportation is an essential part of modern life. People don't travel long distances by boat as much nowadays due to lack of speed, but *shipping* got its name for a reason. We still use ships to move vast quantities of cargo around the world every day, maintaining complex supply chains of raw materials to finished goods. Waterway transportation persists because ships are efficient. Even the most massive of consignments are practically effortless to move once floating on water. Moving one ton of goods the same distance on a boat takes roughly half the amount of energy that it would by train and approximately a fifth of the energy it would take on a truck. In addition, shipping is the primary way goods can move across portions of the globe that aren't connected by land.

Ports are the hubs that connect maritime and overland modes of transportation. In the simplest terms, a port is a place where a ship can dock, but that straightforward function belies the enormous complexity of modern maritime facilities. Ports are found not only in cities along the coast but also in cities along rivers and inland waterways. They usually consist of multiple *terminals* where loading and unloading of goods—or people, in the case of cruise ships—take place. Each terminal is engineered to move a specific type of goods on and off ships quickly and efficiently. *Bulk carriers*, which transport unpackaged cargo like grains and ore, are serviced using large conveyors or bucket cranes.

Tankers, which carry liquids like oil, are filled and drained by massive hoses. Most cargo ships that move packaged goods use CONTAINERS, standardized steel boxes that can be easily transferred between trains, trucks, and other vessels using cranes.

A container terminal is one of the most recognizable parts of a commercial shipping port with its gigantic cranes and colorful stacks of steel boxes. The enormous SHIP-TO-SHORE CRANES usually sit atop rails so they can traverse the length of a cargo ship, loading or unloading containers as quickly as one every two minutes.

Sometimes containers are moved directly between modes of travel (mainly truck, train, or another ship), but often they must be stored in the YARD before their onward vehicle arrives. Containerization of cargo creates a puzzle since only the top container in each stack is accessible. Getting to the bottom of the stack requires relocating everything above. Computerized management systems optimize the placement of each container to reduce the number of movements it takes to deliver them to their destinations.

A wide variety of vehicles are used to handle and move containers within the terminal, and their control is increasingly being automated in modern ports. TERMINAL TRACTORS (known by many names, including *hustlers*) are small semi-trucks that carry containers around the yard. AUTOMATED GUIDED VEHICLES perform the same function but without the human driver. REACH STACKERS and STRADDLE CARRIERS can transport and lift containers to and from the top of a stack. GANTRY CRANES ride over long rows of stacked containers. Instead of hooks, these vehicles all use a device called a *spreader* to lift each container. Every box is equipped with reinforced CORNER CASTINGS. Four TWIST LOCKS engage into oval-shaped holes in each casting. The twist locks rotate 90 degrees, securely connecting the spreader to the container. Ingenious in their simplicity, twist lock mechanisms are installed on ship decks, trucks, and trains, and between each container in a stack. They are responsible for locking millions of hulking steel boxes into place each day.

Although maritime terminology varies regionally and worldwide, the structure that serves as the edge of the terminal is usually called a *wharf* or *quay*. The wharf may include one or more *berths*, which are parking spots for ships. Each berth includes several large BOLLARDS to which the ship's MOORING LINES are attached. *Winches* on the ship keep these lines taut to minimize movement during loading and unloading. In addition, FENDERS along each berth serve as cushions to protect both the wharf and the vessel's hull from damage. Traditionally, old tires were used as fenders, but modern ports use devices specifically designed for the types and sizes of ships serviced.

One of the most critical decisions in the design of a port facility is the largest ship that can be accommodated, called the *design vessel*. Accommodating larger ships makes port facilities more expensive to construct and maintain, but it can bring more traffic and more revenue, so it's a careful balance. The design vessel length determines the length of each berth and the overall size of the harbor. Its BEAM affects the size of the BOOMS on the ship-to-shore cranes used to load and unload, and its DRAFT determines the minimum depth of the harbor below. This depth is maintained by dredging sediment from the waterway floor using excavators or suction pipes. Designers of ships (called *naval architects*) try to make them as large as possible while still fitting into the canals, locks, and ports they will encounter. In fact, many ship types are named for the facility into which they will barely fit; for example, Suezmax ships are the largest vessels capable of transiting the Suez Canal.

Wharves must be robust structures, capable of withstanding wind, waves, tides, currents, and the extreme forces of ships' mooring lines day in, day out. In addition, they must be quite tall to allow enormous ships to dock directly alongside. Many wharves are built on FILL, soil brought to the site and compacted in place to serve as a firm foundation. A RETAINING WALL reinforces the fill while allowing ships access to the edge. When site geology isn't ideal for supporting the weight of port equipment and cargo, a wharf may be supported on PILES. These vertical steel or concrete elements are drilled or driven deep into the underlying soil to keep the wharf from settling or shifting over time.

Waterways use many navigation aids to help mariners safely steer their ships. Floating devices, called BUOYS, delineate navigable waterways and hazards. Just like road signs, they use standardized colors and symbols to communicate rules and information. They are usually held in place with a chain and anchor. The chain has enough slack to absorb shock loads from waves, wind, and current and accommodate changes in level from tides. The anchor can be a heavy weight, called a SINKER, or a device driven or drilled into the underlying soil.

For much of history, it was common for overloaded ships to succumb to high waves and sink. Without regulations, captains were incentivized to carry as much cargo as they thought the vessel could hold, frequently overestimating to the demise of their goods and crews. Over time, insurance companies and the international shipping community formalized requirements for marking the legal load limit on the outside of every ship. This load line, usually denoted as a horizontal line through a circle, will fall below the water if the vessel is overloaded. It is often known as the *Plimsoll line* after the British politician who championed its use. The buoyancy of a ship depends on the temperature of the water and whether it's in seawater or freshwater, so most modern ships have a set of marks that serve as the load line in the various conditions through which they may travel.

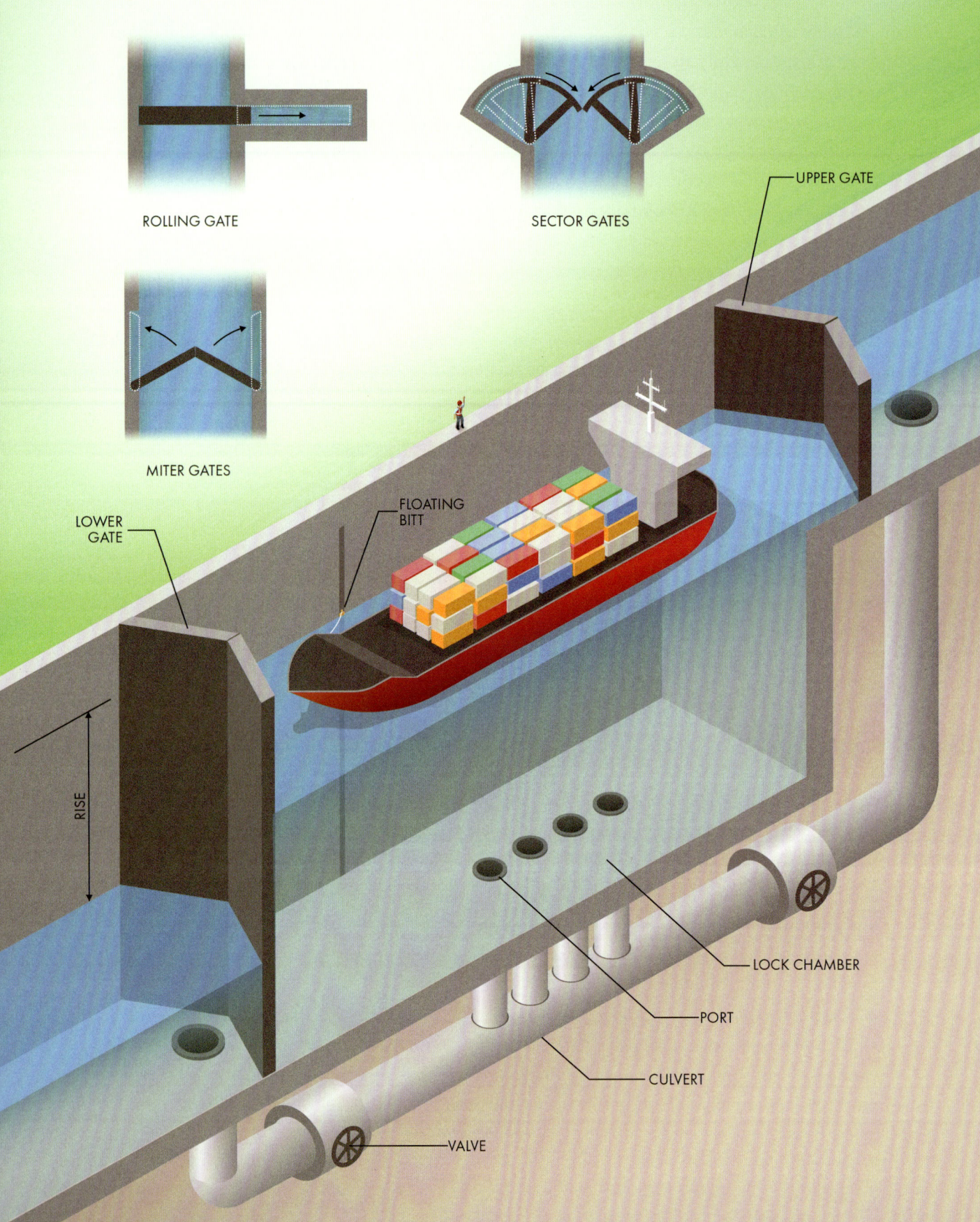

ROLLING GATE

SECTOR GATES

MITER GATES

UPPER GATE

LOWER GATE

FLOATING BITT

RISE

LOCK CHAMBER

PORT

CULVERT

VALVE

Locks

Shipping by waterway has its limitations, namely that not every place is accessible by boat. We've overcome this obstacle somewhat using constructed waterways or canals. The earliest works of written history describe canals and shipping. Even thousands of years ago, humans sought to bring access by boat to areas otherwise inaccessible. Another limitation is more difficult to surmount, however. Water is self-leveling. Unlike roads or rail, you can't lay water on a slope to get up or down a hill. An ideal canal would sit at the same level for its entire length, but in areas with steep terrain, that would require so much excavation that it would be practically impossible. Rather than cut massive canyons to keep canals at a consistent elevation, we move ships up or down to different levels like steps on a staircase using navigation *locks*.

A lock consists of a watertight CHAMBER with large gates at either end. The way a lock works is quite simple. A ship going up enters the mostly empty chamber, and the LOWER GATE is closed. Then water from above is allowed to fill the space, raising the vessel. Once the level in the lock reaches that of the upper canal, the UPPER GATE can be fully opened, and the ship can continue on its way. Going down follows the same steps, but in reverse. A ship enters the filled chamber, the upper gate is closed, and the water in the lock is allowed to drain. Once the level in the chamber matches that of the lower canal, the lower gate can be fully opened, and the

ship can continue onward. It's an entirely reversible lift system that, in its simplest form, requires no external source of power to work, except for the water itself.

Locks on rivers may be combined with a dam that impounds water and discharges floodwaters when needed. Most modern locks that accommodate large ships are built from reinforced concrete. They have walls and a floor, just like a gigantic bathtub. The approaches to the lock are designed to be straight with no crosscurrents so ships can easily line up to enter the chamber. Small locks used for recreational boats can often be self-operated, but large locks on busy waterways have operations staff to lift and lower ships 24 hours per day.

The gates on each side of the LOCK CHAMBER are engineering marvels themselves. Most locks use MITER GATES. They are composed of two leaves, like enormous, hinged doors that close toward the center. Instead of closing in a straight line, the leaves meet at an angle pointing upstream. Pressure from the water above forces the gates tightly closed, keeping them sealed and leak-free during the operation of the lock. In some places, particularly those affected by tides, it is possible for the downstream level to rise above that in the upper canal. In such a situation, miter gates would not function properly. SECTOR GATES provide an alternative to miter gates that can handle water pressure from both directions. Sector gates are shaped like slices of pie that hinge at their points and meet in the center. Some modern locks employ gates that roll open and closed instead of using hinges. ROLLING GATES benefit from sliding into a recess that can be pumped dry for maintenance and repairs (rather than needing to remove each gate entirely).

In all locks, the lower gate is the real workhorse. The upper gate must only be tall enough to allow ships to enter the chamber when filled. The lower gate must hold back water from the very top to the bottom of the chamber. Water pressure increases with depth, so locks with a significant RISE require the lower gate to withstand extreme forces. When a canal needs to ascend a substantial elevation difference, multiple smaller locks in series (called a *flight*) are used rather than a single large lock.

The plumbing required to fill and empty a lock chamber is another essential part of its engineering. Many locks are choke points to waterway traffic, so operators try to minimize the time it takes a ship to get through. Imagine the challenge of filling and draining a gigantic swimming pool 30 or more times per day while people are in the pool. Similarly, you can't just open the upper gate and let the water flow in. For one, the difference in water levels creates so much pressure that opening the gate is practically impossible. More important, the in- or outrush of water would endanger ships transiting through the lock. Instead, most locks use a separate system to fill and drain the chamber. The simplest option is to have a smaller shutter, sometimes called a *paddle*, in each gate that can be opened and closed. Large locks use CULVERTS to move water through PORTS in the sides or bottom of the chamber. Two VALVES control the flow. The lock is filled by opening the valve at the upper gate, and the chamber is drained by opening the one at the lower gate. The ports are carefully designed to move as much water as possible without creating dangerous turbulence, jets, or swells within the chamber that could capsize a ship.

Even with a well-designed filling system, a lock chamber can be a turbulent space. Ships need to be moored in place to avoid collisions with the gates or walls. However, mooring lines can't be attached to the tops of the lock walls. For a ship moving up, they would immediately become slack. For a ship moving down, they could pull the vessel right out of the water! Smaller locks require the boat pilot to take in or pay out the lines as the water level rises or falls. Larger locks use FLOATING BITTS that ride along vertical guides to keep moored vessels in place as they ascend or descend.

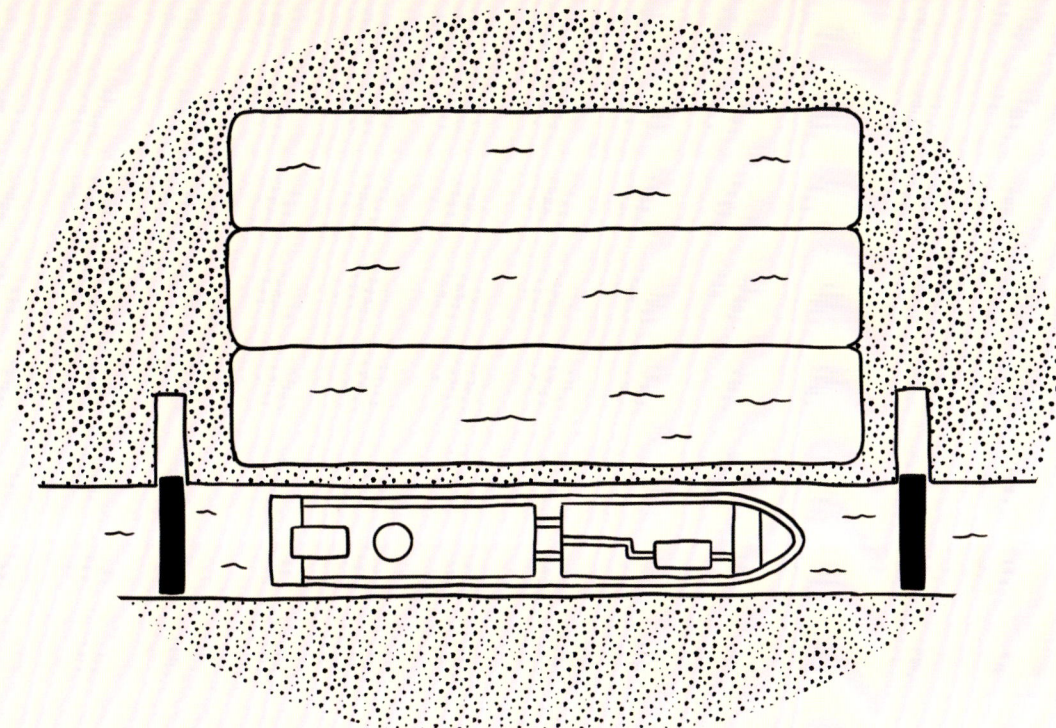

Even though boats can move through locks in both directions, water moves through in only one. Each time the lock is operated, you lose a "lockful" of water downstream. Canals aren't always full of limitless water, and operating locks day in, day out can mean losing millions of liters of water per day. Some facilities use *water-saving basins*, also called *side ponds*, to reduce the loss of water through the locks. The lock chamber drains into the nearby basin rather than releasing the water downstream when lowering a ship. When the time comes to fill the lock, water from the basin is used first to raise the level as far as possible. The remainder of the chamber is supplied from the upper canal.

Without large pumps, the water savings are limited by gravity. A side pond must sit at an elevation between the top and bottom of the lock chamber so water can flow both into and out of it, meaning only around a third of the water can be recycled. However, the savings can be increased with the size and number of basins. For example, the newest locks at the Panama Canal have three basins each, allowing them to use only about 40 percent of the water that would otherwise be required.

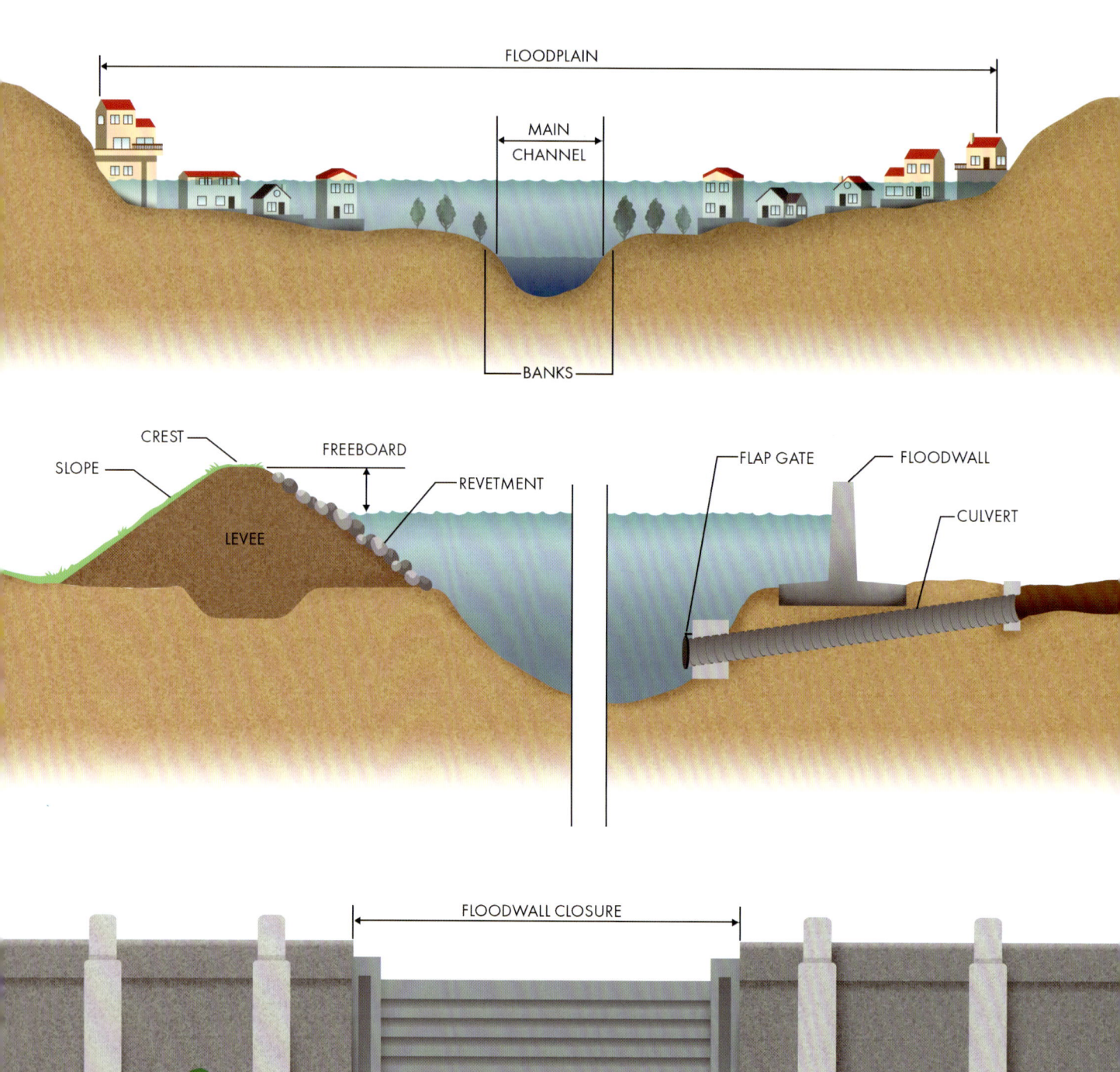

Levees and Floodwalls

Every year floods impact populated areas, costing lives and millions of dollars in damages, devastating communities, and grinding local economies to a halt. If you've ever experienced one yourself, you know how powerless it feels to be up against Mother Nature. We can't change how much it rains, but we have developed ways to manage that water once it reaches the land to limit the danger it poses to lives and property.

Flooding on rivers is particularly challenging to manage because its impacts aren't linear. In the MAIN CHANNEL, where normal river flows occur, a rise in water level creates only a slight increase in the area of inundation. The steep BANKS contain the flow. However, the topography is often wide and flat above the channel banks, ideal for farmland and urban development. When the river's flow overtops the banks, even small increases in water level can create broad areas of inundation. These areas just above the riverbanks are often called FLOODPLAINS because they are so vulnerable to overbanking flows. One structural solution to riverine flooding is to increase the height of the banks to constrain the flow of water away from developed areas.

The most common way to raise the banks of a river is simply to gather nearby soil and pile it into an embankment. These structures, called LEVEES or *dikes*, have been used for centuries to divert and impound water. They are also used along coastal areas to protect against storm surge. Although simple in concept, modern levees rely on advanced engineering to protect low-lying areas from floods. After all, soil is not the strongest of building materials, especially when faced with fast-moving water. Engineers specify the SLOPES and compaction requirements of the levee based on the properties of the soil available for construction.

Rapid flows during floods can create erosion and damage the river side of a levee. The slopes are often planted with grass as the dense root systems protect against erosion. Levees subject to longer-duration floods or high waves may include stone or concrete armoring, called REVETMENTS, for additional protection. Because earthen embankments can deteriorate over time, maintenance is vital. Levees must be kept free of trees and woody vegetation that can be toppled or ripped out during a flood. Burrowing animals must also be discouraged from making a home in levees because their holes can create paths for water to seep through the soil structure.

Although relatively inexpensive and straightforward, levees take up a significant amount of land due to their trapezoidal shape. A more expensive but space-saving alternative is to build a FLOODWALL. These walls are typically made from reinforced concrete and serve the same purpose of raising the banks of a river to keep the flow contained. They are also less susceptible to long-term deterioration because they are built from more resilient materials than compacted soil.

The height of a levee or floodwall is a critical decision. The potential magnitude of flooding is almost limitless. If you can imagine a big storm, you can probably imagine a bigger one, which means flood infrastructure must strike a balance between cost to build and the amount of protection it provides. In the United States, many levees and floodwalls are designed to guard against the *100-year flood*, a somewhat confusing term for a simple concept. Because we have extensive historical records of rainfall worldwide, we can estimate the relationship between any storm's severity and its probability of occurring. The 100-year flood is a reference point on that line: a theoretical storm that has a 1 percent chance of being equaled or exceeded in a given year at a certain location. Although its name implies that it happens only once every hundred years, the 1 percent annual chance equates to a 26 percent chance of such a storm occurring within a 30-year window. Over 50 years, that probability approaches 40 percent, nearly the flip of a coin.

Designing to the 100-year flood is our way of recognizing that it isn't cost-effective to protect against all floods, but we can design our infrastructure to protect against them 99 percent of the time. To set the CREST of a levee or floodwall, engineers use historical flooding records and hydraulic models to estimate how high the 100-year flood would reach along a river. Then they add a little extra, called FREEBOARD, to account for uncertainty and to keep waves from overtopping the structures.

Completely enclosing an area at risk of flooding with a levee or wall isn't always possible. For one, roadways and railroads need a way to cross protected areas. There isn't always enough space or funding to build ramps or bridges up and over each wall, so we occasionally leave a gap, called a CLOSURE, through which a road or railway can pass. Steel gates at each closure must be shut in advance of a flood to complete the perimeter. Of course, closures are only feasible for areas along major river basins where floods arrive gradually with some warning. An open gate can defeat the purpose of a floodwall or levee altogether, so closures can't be used in areas susceptible to flash floods.

In addition, enclosing a low-lying area of the landscape creates a basin that can fill up with water on the wrong side of the wall during storms. Levees need a way to allow drainage to pass through in one direction without allowing the river to backflow into protected areas during floods. Some large-scale systems use pumps to move local drainage out of low-lying areas, but pumps can be expensive. Pipes, called CULVERTS, can pass through levees and floodwalls, or their foundations, to allow drainage of enclosed areas to occur by gravity. These culverts are equipped with gates (which must be manually closed during floods) or devices that defend against backflow automatically, called *check valves*. FLAP GATES are a common type of check valve that seal shut against water pressure coming from the opposite direction.

Although levees protect low-lying areas from flooding, they can create new issues too. Since levees constrain the power of a river to a smaller space, the water flows higher and faster than it would have without such structures, potentially exacerbating flood impacts farther downstream. Even with excellent engineering, our ability to "control" Mother Nature is usually tenuous. Flood control infrastructure is vital in developed areas, but it must be combined with management and respect for rivers' natural floodplains.

KEEP AN EYE OUT

A common technique for fighting floods is to use stacked bags of sand to impound or divert water. With only a small crew of workers, sandbags can be added to the top of a levee to increase its height or around an unprotected structure to keep floodwaters out. Each bag is typically filled about half full, so it conforms readily with its neighbors without leaving significant gaps. A small trench in the center helps them key into the foundation to withstand the pressure of floodwaters. Bags are stacked in a pyramid shape with a bottom width about three times the height of the berm. A plastic sheet can be added to the upstream face to make the barrier more impermeable.

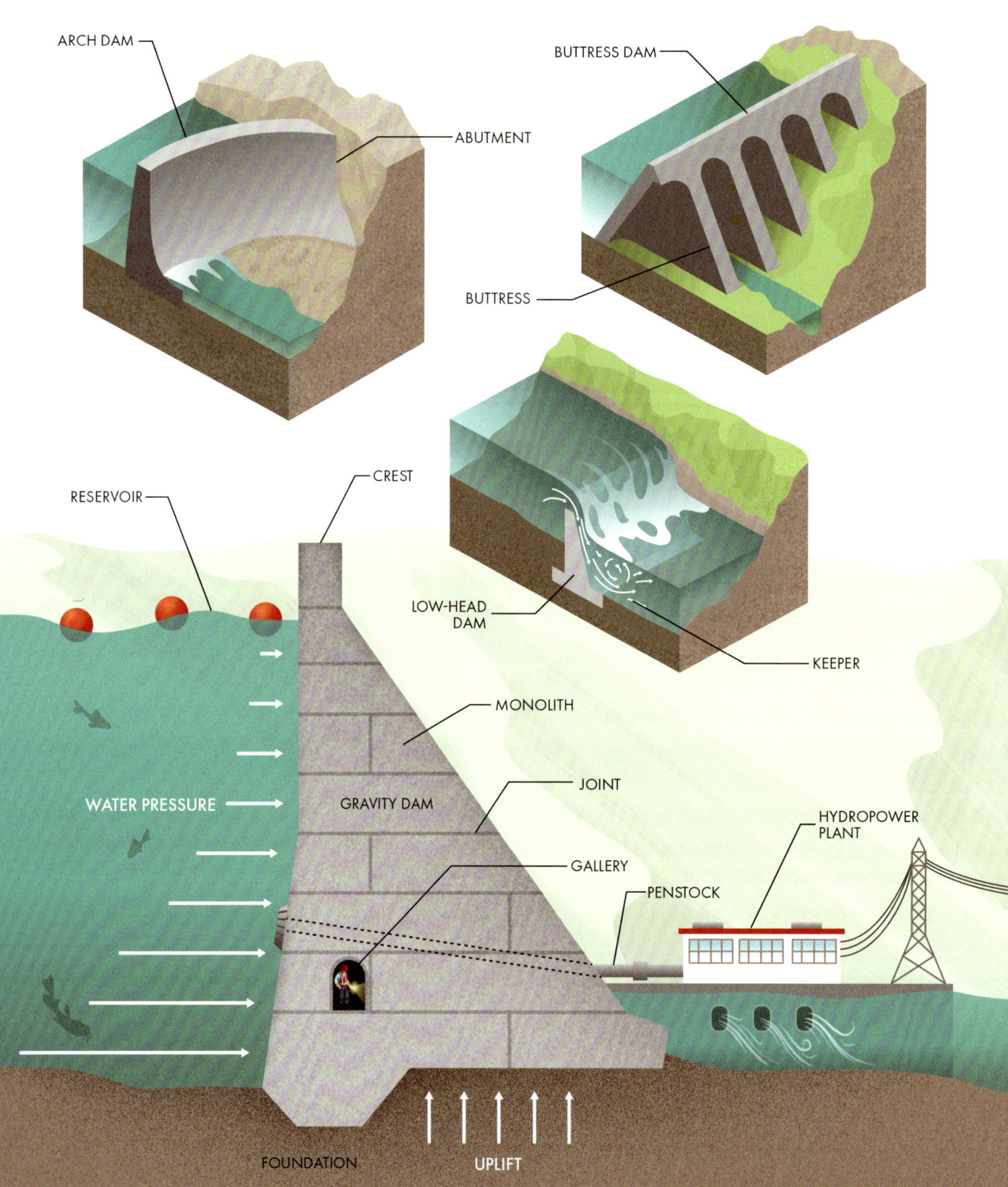

ARCH DAM

ABUTMENT

BUTTRESS DAM

BUTTRESS

RESERVOIR

CREST

LOW-HEAD
DAM

KEEPER

MONOLITH

WATER PRESSURE

GRAVITY DAM

JOINT

HYDROPOWER
PLANT

GALLERY

PENSTOCK

FOUNDATION

UPLIFT

Concrete Dams

Water is one of Earth's most essential resources, but the hydrologic cycle comes with tremendous variability. From droughts to floods and everything in between, achieving a consistent water supply can be a significant challenge. We can't control how much it rains or how often, but we can develop storage to smooth out the highs and lows of inflow throughout each year. Building a dam across a river valley creates a *reservoir*, a place where water is stored and can be used over time for irrigating crops, providing water to cities, or generating electricity.

A reservoir can also be kept empty in anticipation of severe weather, allowing a dam to hold back flood waters and release them gradually, reducing the damage they inflict downstream. (Spillways used to discharge water are discussed in a later section.) Many large dams serve multiple purposes at once using different zones within the reservoir, called *pools*. One pool can be kept full to be used for hydropower or water supply, and one above kept empty to be used for storage in the event of a flood. If a dam is used to generate electricity, the hydropower plant that houses turbines and other equipment is often visible downstream. If the plant isn't connected to the dam, you might also see the large-diameter pipelines that deliver water to the turbines, called PENSTOCKS.

A dam can be built from many different materials, but many of the largest and most iconic structures are made from concrete. (The following section describes dams constructed from earth and rock.) Concrete is strong and durable, allowing a dam to withstand the tremendous pressure of water in a reservoir. Unlike many large structures where loads are vertical from the force of gravity, the most significant forces on a dam are horizontal. As the depth of a reservoir increases, so does the PRESSURE it exerts on the upstream face of a dam. Water can also leak through pores and cracks in a dam's FOUNDATION, creating pressure on the bottom of the structure, called UPLIFT. The critical job of withstanding such pressure is a major factor in each dam's design and physical appearance.

GRAVITY DAMS oppose the force of impounded water simply with their weight. Concrete is quite heavy, and with enough mass, a structure can be stable enough to avoid overturning or sliding from horizontal forces. Gravity dams are usually broad at their base, where water pressure is highest. They taper up to a narrow CREST, sometimes just wide enough to drive a vehicle over the top, giving them a characteristic slope on the downstream side. Similarly, BUTTRESS DAMS transfer the forces from the reservoir into the foundation using triangular BUTTRESSES. The water pressure still works to push the dam horizontally, but the sloped upstream face takes advantage of the water's weight for stability as well. Buttress dams require less concrete to construct, but they also require more labor to form the intricate shapes needed for stability, so they are generally not economical in modern times.

Unlike gravity and buttress dams, ARCH DAMS transfer much of the force from impounded water into the ABUTMENTS on each side of the dam instead of the foundation. Like arch bridges (described in Chapter 4), arch dams take advantage of geometry to span a gap. Because they don't rely as much on their own weight, arch dams need less concrete and thus can be more economical to construct. However, they can be located only on sites with advantageous geology since the abutments need to resist most of the reservoir forces attempting to push the structure downstream. Thus, arch dams are most often found in narrow, rocky valleys. Some buttress dams are designed as *multiple arch dams*, where each smaller arch is supported by a buttress instead of a single arch spanning the entire valley.

Concrete dams are not constructed as one solid block—concrete shrinks as it cures from a liquid into a solid, potentially leading to cracks. In addition, changes in temperature cause concrete to expand and contract over the course of a year, which can also cause cracking. Cracks in a sidewalk or driveway may be okay, but in a dam, they can become leaks that weaken and damage the structure. Concrete dams are constructed in smaller blocks, called MONOLITHS, with horizontal and vertical JOINTS to provide freedom of movement, reducing the likelihood of cracking. Unlike random cracks that could form in a solid concrete structure, joints are easy to seal

against leakage using embedded water stops and sealants. Although you can't see them from the outside, many concrete dams have internal tunnels, called GALLERIES, to collect any leaking water and allow engineers to monitor the structure's integrity from the inside. Galleries also provide a location for drains that relieve pressure within a dam's foundation.

Another type of concrete structure, usually called a LOW-HEAD DAM, is not used to create storage but simply to raise the level in a river or stream. The depth in a natural watercourse varies over time and can be quite shallow for long stretches. Low-head dams impound a small volume of water, artificially raising the level to make a channel more navigable by boats, increase the depth at intakes for water supply and irrigation, or create a drop for powering turbines or waterwheels. Low-head dams are often called *weirs* because water flows over the top (rather than through a gate or outlet). This overflow can create a significant hazard for swimmers and boaters.

As the jet of water (called the *nappe*) flows over a low-head dam and plunges into the river below, it can create an area of recirculation immediately downstream of the dam. This area is sometimes called a KEEPER because it can trap objects, debris, and even people. With strong hydraulic forces, the hard surfaces of the dam, disorienting turbulence, and submerged debris, the low-head dam has been known as the perfect drowning machine. Many were constructed long ago when mills and factories relied on waterpower to drive equipment, and safety was low on the list of priorities. Many cities have removed them or converted them into recreational features, restoring aquatic ecosystems and drawing outside tourism. If you're swimming or paddling on a river with a low-head dam, don't underestimate the danger of these seemingly innocuous structures.

Dams are high-risk structures. Since a failure could send a severe flood wave downstream and threaten populated areas, most large dams have comprehensive monitoring plans to keep them safe. In addition to regular inspections by engineers, many dams have instruments that monitor the structures' integrity. Instrumentation can measure water pressure within a dam or its foundation, settlement or movement, water flow in drains, and even the temperature of concrete over time. Devices can be sensitive enough to see a dam subtly expanding in size from heat on a sunny day. Many dams are also equipped with surveying monuments whose location can be accurately tracked over time by precision measurement equipment. All the data from a dam's instrumentation provides early warning of conditions that could contribute to a failure, allowing engineers to evaluate and repair problems before they lead to hazardous conditions.

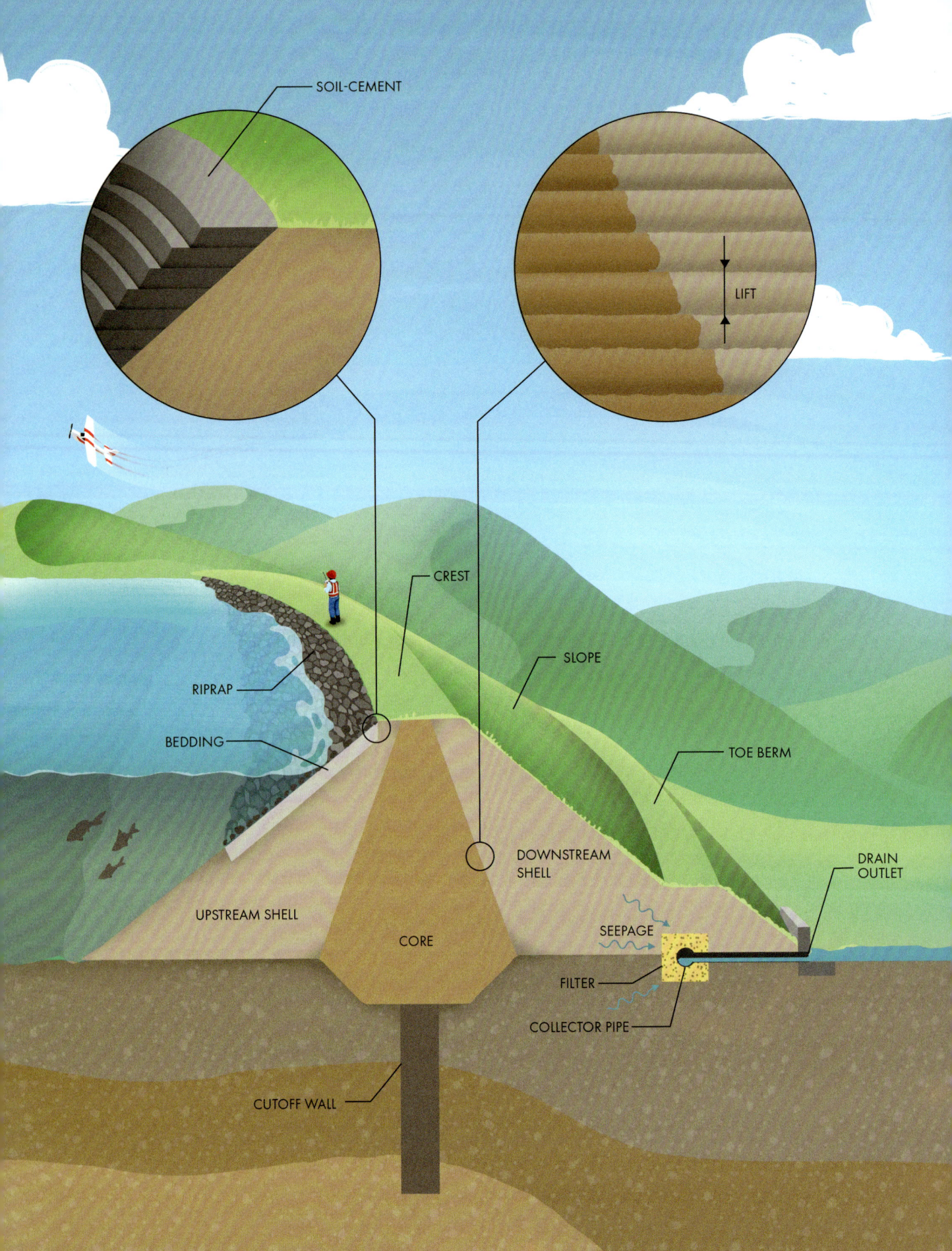

SOIL-CEMENT

LIFT

CREST

SLOPE

RIPRAP

TOE BERM

BEDDING

DOWNSTREAM
SHELL

DRAIN
OUTLET

UPSTREAM SHELL

CORE

SEEPAGE

FILTER

COLLECTOR PIPE

CUTOFF WALL

Embankment Dams

Although the archetypal dam is a concrete structure, most dams worldwide are constructed using earth or rock. Unlike concrete dams that usually require specific geology and a nearby source of constituent materials (mainly cement and aggregates), *embankment dams* can be constructed in nearly any location. If there are two materials in abundant supply on Earth, they're soil and rock. An embankment dam is not just a pile of dirt placed across a river valley, however. Using such primitive materials to impound enormous volumes of water safely is a complex engineering challenge, and attentive observers can notice many of the intricacies of embankment dam design.

Embankment dams can be constructed out of soil, called *earthfill*, or out of stones or gravel, called *rockfill*. Both types of material behave much differently than concrete. Because they are granular substances composed of individual particles, earthfill and rockfill are naturally unstable. Gravity is always trying to pull them apart, and the only force holding an embankment together is the friction between individual grains or rocks. Large embankments that can stand tall over the long term and resist pressure from a reservoir must have gentle SLOPES on the upstream and downstream sides. The necessary slope depends on the properties of the material being used, but most earthfill dams have slopes with widths that are around three times their height. The slopes of rockfill structures can

be steeper but rarely less than a two-to-one ratio of width to height. That means both earthfill and rockfill embankments have broad footprints that taper as their height decreases toward each end. Many also feature a TOE BERM, an area of additional fill along the bottom of one or both slopes to stabilize the structure further.

Soil and rock can't simply be dumped into place to create a dam. Granular materials settle and compress over time, and the height of the pile magnifies this effect. We don't want dams shrinking after they're built, so fill must be compacted and densified during placement to create a firm and stable structure. Compaction speeds up the settlement process, so it mostly happens during construction instead of afterward. If soil is compacted to its maximum density, that means it won't settle further over time. Modern construction equipment can compact a layer of soil up to around 30 centimeters (1 foot) thick at a time. Rolling over thicker layers will only densify the surface, leaving the underlying ground loose. So, embankments are constructed slowly from the bottom up in individual layers called LIFTS.

Rockfill and most types of earthfill are permeable materials that can allow water to flow right through (a phenomenon called SEEPAGE). Unlike concrete dams, which accomplish both stability and water tightness with one material, embankment dams often require additional features to hold a reservoir back. Most earthfill dams have different zones of material. The CORE is constructed using clay soils that are highly impermeable to seepage. Depending on the site's geology, finding clay in sufficient volume that meets strict specifications for water tightness can be a challenge. The core is typically the costliest part of an embankment project, so its dimensions are engineered to be only as large as necessary. The outer SHELLS have less stringent specifications because they provide only stability and don't need to be so watertight.

Rockfill dams, being much more porous than earthfill structures, usually include a barrier of concrete, asphalt, or clay in the core or along the upstream slope to make the embankment impervious to seepage. In addition, although not visible from the outside, many embankment dams feature some kind of CUTOFF WALL into the foundation. This wall is often constructed from concrete or a clay slurry to close any paths that seepage could take from the reservoir through the dam's foundation.

The repeated force of waves crashing against a vulnerable earthen structure can create erosion and deterioration. Therefore, nearly all large earthfill dams will have some kind of armoring on the upstream face to protect against long-term damage from waves. Often this armoring consists of a thick layer of rocks called RIPRAP. A layer of smaller gravel called BEDDING sits between the dam and larger rocks to prevent soil from washing out from under the riprap. Alternatively, many dams use a mixture of onsite soil and cement, forming an inexpensive but durable

armoring called SOIL-CEMENT. It is often placed in lifts along the upstream face of embankments, forming a characteristic stair-stepped appearance.

Beyond the limits of any armoring, embankment dams are usually covered in grass to protect against erosion from rainfall runoff. With their gentle grassy slopes, many embankment dams appear at first glance to be a natural part of the landscape. If you can't see the reservoir on the other side, you might not even realize a dam is there at all, except for the perfectly level CREST, which often gives it away.

All dams leak at least a little. Achieving perfect water tightness for such massive structures usually isn't worth the cost. Instead, engineers make sure that leaks don't cause problems through the use of drains. Most drains consist of two parts: the FILTER keeps seepage from washing away soil particles using layers of gravel or sand, and perforated COLLECTOR PIPES within the filter gather and discharge any water that finds its way into the drain so that it can't build up pressure. If you see small pipes exiting the downstream side of the dam, they are often OUTLETS to the structure's internal drainage system.

Some dams are not constructed across a stream or river, but rather in upland areas nearby. *Off-channel reservoirs* are those created by building a circular dam to contain the stored water fully. They must be filled using pumps from a nearby source of water (usually a river). They are often more expensive because the dam must surround the entire perimeter. Still, off-channel reservoirs are less disruptive to the natural environment because they don't create a barrier across a river, and they can be constructed on less sensitive sites.

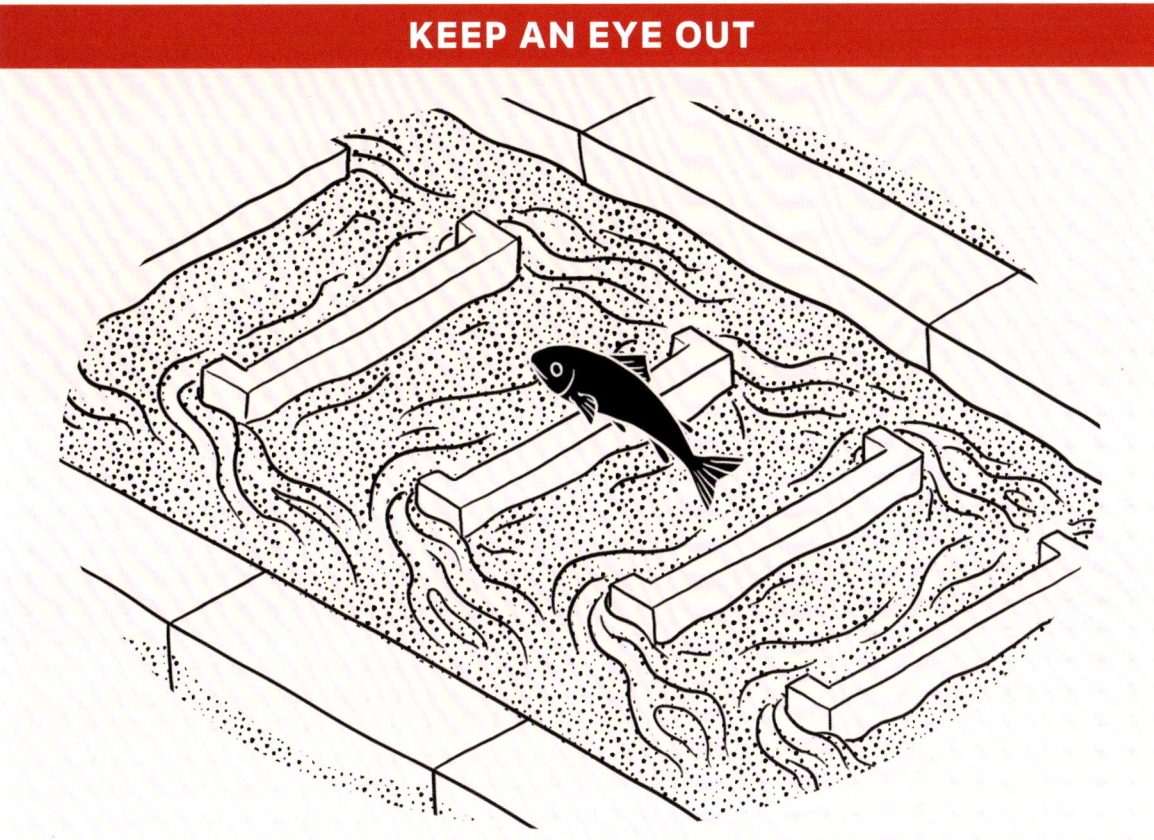

Although they are vital to humans by storing water, preventing floods, and providing a renewable source of electricity, dams can severely disrupt the natural environment. Many were constructed before strong environmental regulations were in place, leading to unmitigated damage of aquatic ecosystems and natural hydrologic processes. One of the most significant problems they can create is blocking rivers that serve as pathways for migratory fish species. To address this issue, some dams and other artificial barriers in rivers are equipped with *fishways* (also known as *fish ladders*) to provide a bypass to the other side. Although various designs are used, most feature a series of pools with low jumps or cascades through which fish can leap. Designing a structure that mimics the flow of a natural river while ascending or descending a sizeable vertical distance is a challenge, and some configurations are more effective than others. However, biologists and engineers continue to work together to reduce the impacts dams have on the natural environment.

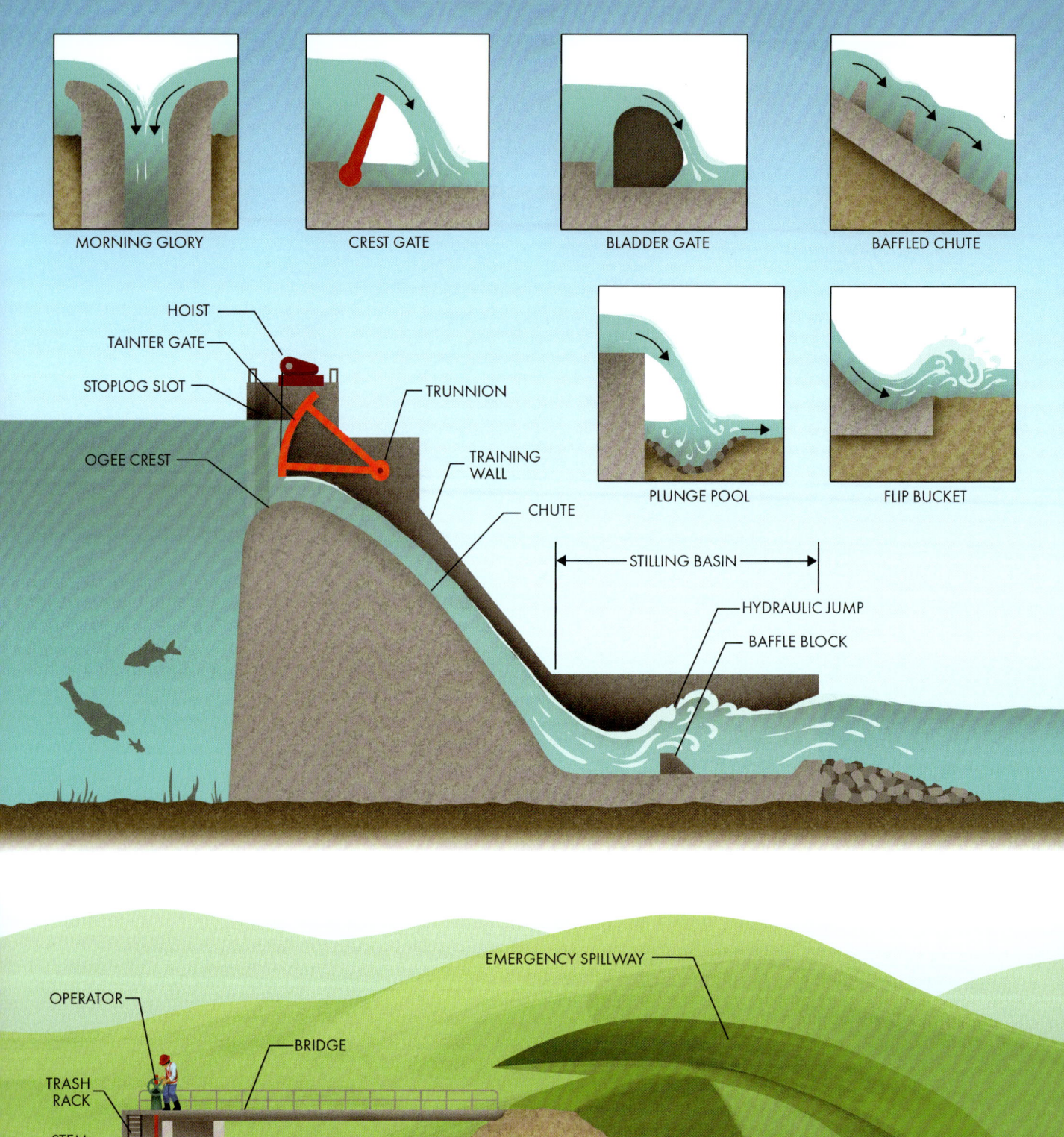

MORNING GLORY

CREST GATE

BLADDER GATE

BAFFLED CHUTE

PLUNGE POOL

FLIP BUCKET

HOIST

TAINTER GATE

STOPLOG SLOT

TRUNNION

TRAINING WALL

OGEE CREST

CHUTE

STILLING BASIN

HYDRAULIC JUMP

BAFFLE BLOCK

EMERGENCY SPILLWAY

OPERATOR

BRIDGE

TRASH RACK

STEM

TOWER

LEAF

CONDUIT

IMPACT BASIN

Spillways and Outlet Works

Although dams are intended for storing water, they need a way to release that water as well, whether it's because the water is needed or to protect the dam from becoming overfull. There are a variety of different structures that can be used to discharge water from a dam safely, depending on the purpose and capacity. Releasing water is a dynamic process, so spillways and outlet works are often the most complex components of a dam.

Although the terminology can vary, *outlet works* are generally the facilities that release water from a reservoir to meet downstream needs. Some reservoir outlets deliver water to a pump station that sends it through pipelines for irrigation or to a treatment plant for drinking water in populated areas. Others provide water to the penstocks of hydropower plants. Still others release water back into the river so it can be withdrawn downstream or be used to maintain the aquatic ecosystems below the dam.

Outlet works are sometimes tricky to spot because they may be fully or partially submerged below the reservoir. They are usually located near the center of a dam where the water is deepest. For concrete dams with vertical upstream faces, the outlet works may be inside the dam itself. Because embankment dams slope away from the center, outlet works are often installed as a separate TOWER farther into the reservoir. A BRIDGE usually connects the tower to the dam's crest to provide access for personnel and vehicles.

The primary features of outlets are the gates and valves that control the flow of water. Before reaching them, usually water must first pass through a TRASH RACK that prevents debris from entering any facilities where it could cause damage. The trash racks on pump station and hydropower intakes often use fine screens to keep fish from being drawn in.

Many types of gates and valves can control water through an outlet. A gate that's stuck open or closed can cause severe consequences, so most outlets use a series of flow controls to provide redundancy and facilitate regular maintenance. Most outlets deliver water through the dam inside a large CONDUIT made from reinforced concrete or steel. One common type of gate used on outlets is a *slide gate*, which consists of a metal LEAF that can be moved up or down across an opening. A STEM connects the leaf to an OPERATOR, often using a motor to lift or lower the assembly. The quality and temperature of water in a reservoir can vary depending on the depth below the surface, so outlet towers often feature multiple gates at different levels, allowing operators to select the elevation at which water is drawn.

One of the greatest risks to dams are floods. Building a dam so tall that it can store the most extreme volumes of floodwater possible isn't practical. On the other hand, a reservoir must never be allowed to overtop a dam because the water will erode and damage the structure and its foundation. So, all dams are designed with at least one spillway, a structure that can safely discharge water downstream when the reservoir is already full.

Because of the inflow variability, many large dams have two or more spillways. The smaller one is called the *principal* or *service spillway* that passes normal inflows when the reservoir is full. The other is called the *auxiliary* or EMERGENCY SPILLWAY that engages only during extreme events. Depending on the design, the auxiliary spillway may flow for only a few scary moments in a dam's entire lifetime, so it can be as simple as an excavated channel cut through an abutment. Sometimes an entire section of the dam is armored to allow it to serve as the spillway, called *overtopping protection.*

Uncontrolled spillways regulate the reservoir level using a *weir*, a structure that allows water to pass over its fixed crest. The volume of water discharged is strictly related to the level of the reservoir and the spillway's size and shape. Many uncontrolled spillways use a curved profile over the crest, called an OGEE, which increases the volume that can be discharged for a given length and flow depth. Some dams use a type of circular weir, called a MORNING GLORY, that discharges into a conduit. This type of spillway is used in narrow canyons where there isn't room for a more conventional overflow.

Controlled spillways use gates to manage discharges. Gates add complexity to a spillway, but they can also reduce its cost by providing flexibility in discharge capacity, allowing for a smaller overall structure. TAINTER GATES have long arms and a curved face, and they pivot around a type of hinge called a TRUNNION. A HOIST above the gate lifts the structure via chains or wire ropes to allow water to flow underneath. CREST GATES pivot at their bottom and are often operated using hydraulic cylinders. Some gates even rely on large rubber BLADDERS that can be inflated with compressed air or water to raise and lower. All gates require regular inspection and maintenance, so most spillways have STOPLOG SLOTS installed upstream. Stoplogs are steel beams that can be installed in the slots using a crane to block the water so that a gate can be isolated for maintenance (called *dewatering*).

As water passes through a spillway or outlet, it drops in elevation between the reservoir and natural watercourse downstream, picking up speed as it falls. In an open channel spillway, the water travels down the CHUTE, and TRAINING WALLS keep the flow contained. Fast-moving water has destructive power that can erode and damage a dam if not carefully controlled, which means both spillways and outlet works need a way to dissipate hydraulic energy and slow down the flow before releasing it into a natural watercourse.

Many types of energy dissipation structures are used on spillways and outlet works. Flows traveling in a conduit may use an IMPACT BASIN, which crashes the water against a solid concrete wall. BAFFLED CHUTES use blocks to slow down the water as it travels downward. PLUNGE POOLS allow water to fall into a large, armored hole before leaving in the downstream channel. Larger spillways sometimes feature a FLIP BUCKET at the end of the chute to launch the flow into the air where it breaks into a fine spray. Finally, many spillways use a structure called a STILLING BASIN to protect the dam's foundation from erosion. Stilling basins rely on a phenomenon called a HYDRAULIC JUMP that occurs when fast-flowing water transitions into a slower stream. Most stilling basins use different combinations of BAFFLE BLOCKS to force a hydraulic jump to form. The turbulent jump stays within the stilling basin, allowing a smooth, tranquil flow to leave downstream and minimizing the potential for erosion that otherwise would threaten the structure's integrity.

The flow of water over a weir is related to both the height of the water over the weir's crest and the total length. A typical spillway design goal is to minimize the size (and thus, construction cost) without reducing the amount of water it can discharge. One clever engineering strategy is to fold a weir into a zig-zag shape, providing more length within a smaller footprint. This configuration is often used to increase the capacity of a spillway. Using a folded shape can also make it possible to raise a dam's level (creating more storage) without sacrificing capacity. Weirs that use trapezoidal or triangular shapes are called *labyrinth weirs*, and those that use square cycles are called *piano key weirs*.

7

MUNICIPAL WATER AND WASTEWATER

Introduction

Water is a fundamental human need, but its cleanliness is just as important. Even before the advent of modern municipal engineering, many civilizations had developed strategies for delivering fresh water to urban areas and removing wastewater to prevent it from contaminating water sources. In the 19th century, as cities worldwide began to grow in population and density, the threat to public health from waterborne diseases became more menacing and insidious. The science of sanitation developed as a necessity to keep city-dwellers safe from plague and pestilence. Now nearly all cities and towns have complex systems for delivering abundant and clean water to their citizens and disposing of sewage. Although easy to take for granted, the development and maintenance of municipal water and wastewater systems are massive undertakings, requiring a lot of infrastructure. Much of the pipes and valves in cities are buried below the ground, but you can observe many facilities and equipment if you know where to look.

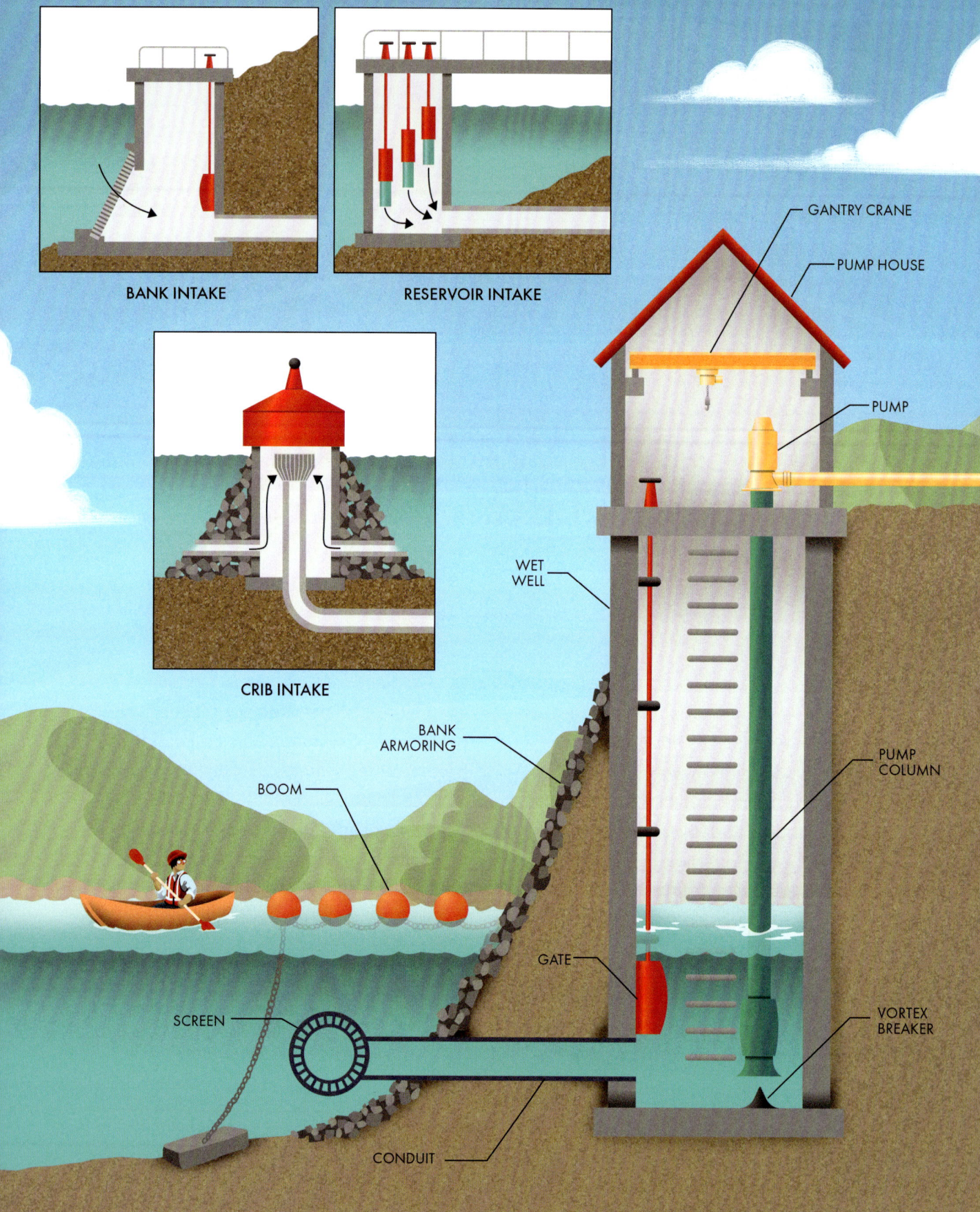

BANK INTAKE

RESERVOIR INTAKE

CRIB INTAKE

GANTRY CRANE

PUMP HOUSE

PUMP

WET WELL

PUMP COLUMN

BANK ARMORING

BOOM

GATE

VORTEX BREAKER

SCREEN

CONDUIT

Intakes and Pumping Stations

Much of the water we rely on for drinking, cleaning, and irrigation of crops starts its journey in a river, stream, creek, lake, or reservoir. These sources are collectively known as *surface waters* (as opposed to groundwater resources, which are discussed in the next section). Collecting water from a river or lake might seem straightforward; however, many engineering challenges are associated with transitioning the flow from a surface water source into a pipeline or *aqueduct* to be delivered to its destination. *Intake structures* perform this critical task. They can be associated with an impoundment or diversion (for example, at a dam), but intakes are often standalone structures, and you'll see them near riverbanks, lakes, or reservoirs if you keep an eye out.

Lake or RESERVOIR INTAKES often consist of large concrete or masonry towers (as discussed in Chapter 6). Complicating matters, a structure considered an *outlet* at a dam might also serve as an *intake* for a pumping station or aqueduct. Older structures called CRIB INTAKES were constructed onshore, floated into place, and then filled with rubble. A central shaft carries water by gravity through the intake to a tunnel below the lake where it can be pumped to treatment and distribution facilities on the shore.

Although complete removal of contaminants and sediment is usually performed later, the engineering of intakes includes making sure that the source water entering a pipe is as clean as possible to reduce the burden on downstream water treatment plants. This untreated water is often known as *raw water*. In reservoirs and lakes, the volume of suspended sediments, the quantity of microorganisms like plankton and algae, and even the temperature of the water can vary significantly with depth. Thus, most intake structures on lakes and reservoirs have openings (or *ports*) at multiple levels so that operators can select the ideal mixture of water from the depths within the lake or reservoir. Gates on the various openings can be opened or closed depending on the conditions within the source water and the needs downstream.

Intakes on rivers contend with a different set of challenges. First, the level in a river can vary significantly. In addition, river intakes must contend with the fact that streams are dynamic systems. Floods can move massive quantities of sediment, changing the location and shape of banks and even altering the course of a river altogether. River intakes are almost always located either on a straight section of the channel or on the outside of a bend. Sediment tends to deposit on the inside of curves where flows are slower, so engineers avoid those locations where an intake could be clogged more readily. BANK INTAKES are installed on a riverbank to allow water to flow laterally into the structure. However, the deepest part of a natural channel (called the *thalweg*) is often in the center, so bank intakes often require dredging of the riverbed to allow water to flow when the river level is low. This dredging is not only disruptive to the sensitive environment of a river, but it must be performed regularly as sediments in the river deposit over time.

One solution to the challenges of varying water levels and sediment buildup is constructing a small *weir* downstream. Such a structure raises the water level in the river while also slowing down the flow so that sediment can settle out. However, weirs create an impediment to navigation and migratory wildlife, and they can be quite dangerous (as discussed in Chapter 6), so their use has fallen out of favor. Modern intake structures on rivers make use of careful siting to avoid problems with sedimentation and lower water levels while minimizing environmental impacts. One alternative to bank intakes involves running a CONDUIT from a deeper section of the channel to the shore, often performed using tunneling to avoid the need for trenching in the natural riverbank. A SCREEN on the end of the conduit prevents fish or debris from entering the line, and a GATE controls the flow.

Unless the final destination of the raw water is well below the source, most intakes will be accompanied by a *pumping station*, which lifts the water from its source into a pipeline or aqueduct. PUMPS are often installed directly above or adjacent to the intake structure, sometimes within a building called a PUMP HOUSE. These structures may be recognizable by built-in GANTRY CRANES that allow equipment to be serviced or replaced when needed.

At a pumping station, water flows into ports, through a conduit or tunnel, then into a structure called a *sump* (or WET WELL), which creates enough volume and depth for pumps to operate. Sumps must be designed to create ideal flow conditions to avoid inefficiencies and damage to the pump. Turbulent and swirling flow in a sump can lead to a *vortex*, just like what happens when you drain a bathtub. If a vortex is allowed to enter the mouth of the PUMP COLUMN, the air will reduce its efficiency and may even cause it to fail. VORTEX BREAKERS are sometimes installed within sumps to keep the flow from swirling as it is drawn into the pump.

Intakes can represent a severe hazard to swimmers and boaters in rivers and lakes because of the submerged structures and fast-flowing water. Where public safety may be at risk, owners of intakes often install floating BOOMS to warn people of the potential danger. These booms consist of brightly colored floating elements linked together with chains. They are anchored to the river or lakebed to create an exclusion zone around a dangerous structure. Some booms are even designed to be strong enough to hold back debris, floating trees, and ice that could damage intake structures. In addition, when an intake or pumping station must be located near the bank, ARMORING (such as *rock riprap*) is usually installed to protect against erosion that could threaten the structure.

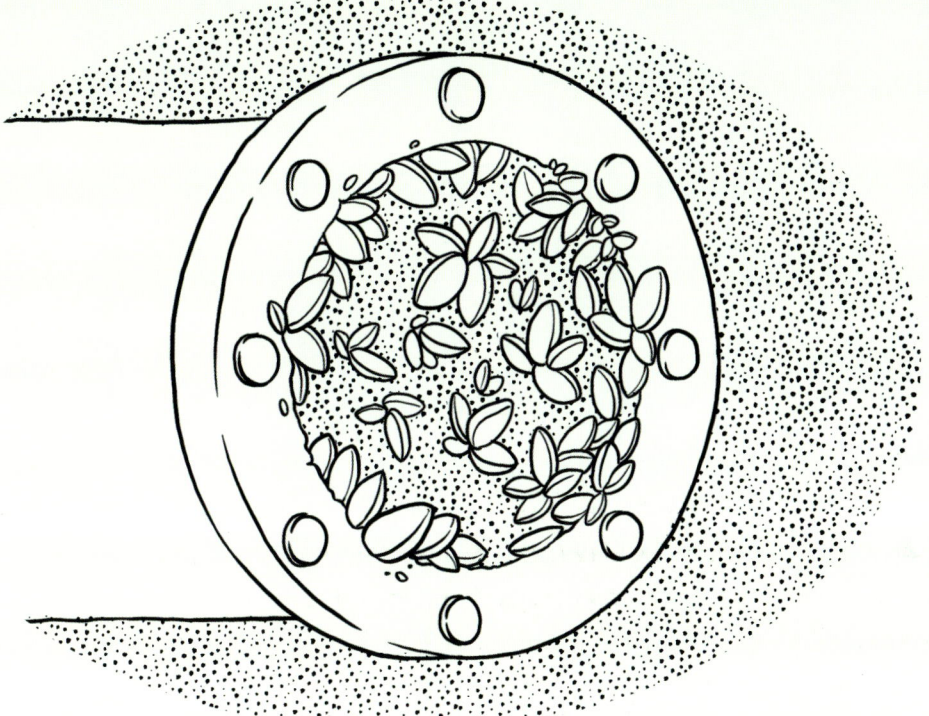

Because intake structures are often installed in natural rivers and lakes, they must contend with aquatic wildlife. Certain types of organisms—including mussels, snails, and clams—can cling to water infrastructure, clogging intakes and reducing the efficiency of pipelines as they accumulate (a process known as *biofouling*). Utilities often employ anti-fouling coatings that discourage animals from attaching or make them easier to remove. However, those coatings must be reapplied regularly, requiring costly shutdowns. In many cases, the most effective solution to biofouling is mechanical cleaning (in other words, scraping the organisms off). Diving teams can clean accessible structures like screens, but pipelines are often cleaned with a cylindrical device, called a *pig*, drawn through the conduit. Many of the most troublesome species are not native to the affected water bodies and thus have less competition to survive, allowing their populations to expand rapidly. One of the most important ways to combat biofouling is to prevent these invasive species from spreading to new water bodies in the first place, so many states have laws requiring boats to be cleaned, drained, and dried before entering a river or lake.

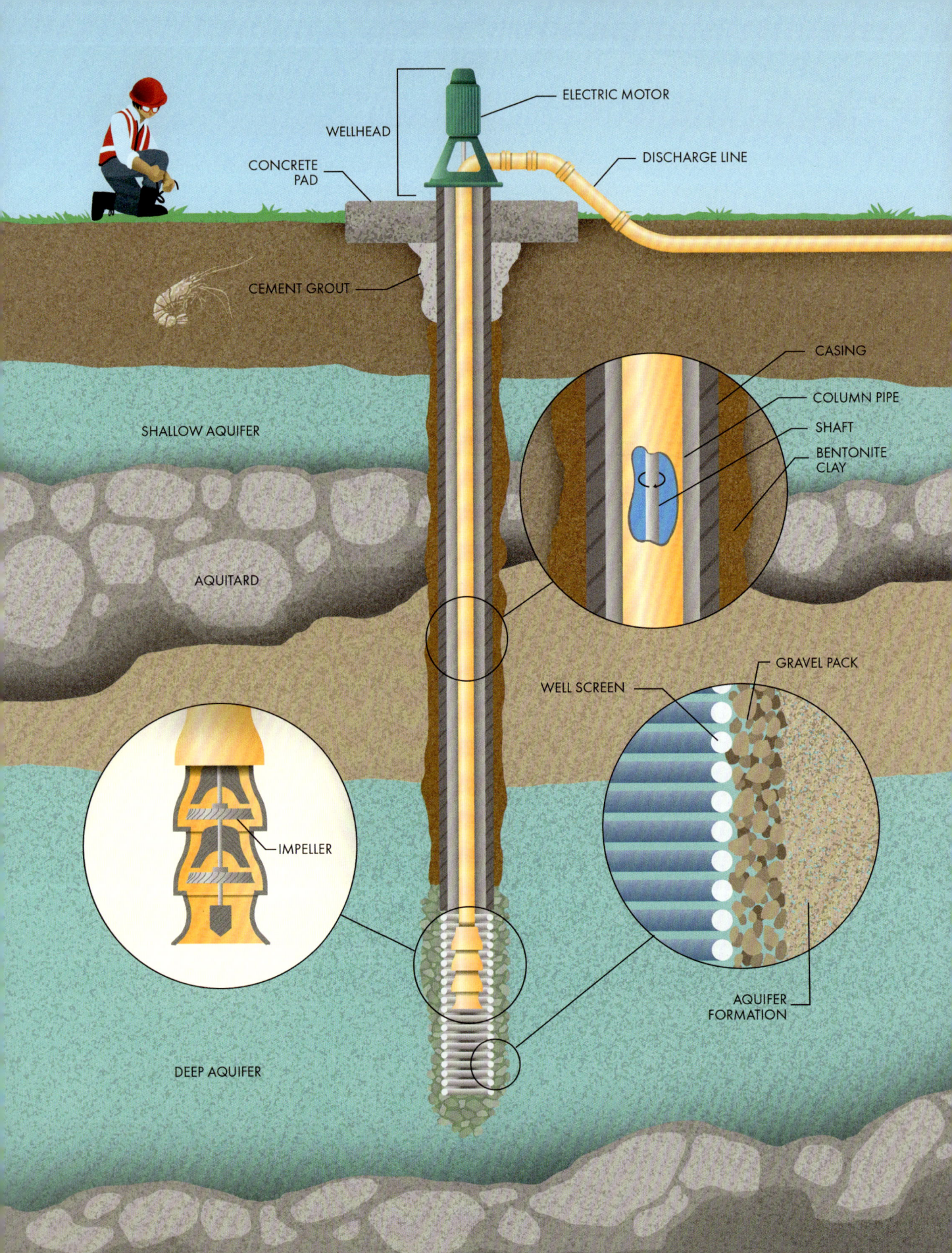

ELECTRIC MOTOR

WELLHEAD

CONCRETE PAD

DISCHARGE LINE

CEMENT GROUT

CASING

COLUMN PIPE

SHAFT

BENTONITE CLAY

SHALLOW AQUIFER

AQUITARD

GRAVEL PACK

WELL SCREEN

IMPELLER

AQUIFER FORMATION

DEEP AQUIFER

Wells

Not all water that falls as precipitation runs off into lakes and rivers. Some seeps down into the ground through the spaces between soil and rock particles. Sometimes this groundwater reaches a less permeable geologic layer (called an AQUITARD) and cannot continue downward. Over long periods of time, infiltrating water can accumulate into vast underground resources called AQUIFERS. A common misconception is that groundwater is stored in open areas like underground rivers or lakes. Although they exist in some locations, large underground caverns are relatively rare. Nearly all groundwater aquifers are geologic formations of sand, gravel, or rock that are saturated with water, just like a sponge. Extracting this groundwater for use by humans is the job of a *well*. At their simplest, wells are just holes into which groundwater can seep from the surrounding soil. However, modern wells utilize sophisticated engineering to provide a reliable and long-lasting source of fresh water. Farms use them for irrigation. Rural homes and businesses often rely on them when a connection to a municipal distribution system is unavailable. And many large cities utilize groundwater as a primary source of fresh water for their populations.

The availability of groundwater varies significantly around the world. Nearly all places have layers of saturated soil or rock below the ground. Still, the volume of water, its quality, and the ease of extracting it to the surface depend primarily on the local geology. Groundwater is also connected to the rest of the hydrologic system, so groundwater withdrawals may impact the volume and quality of surface water resources. Unfortunately, we can't see below the ground, and methods for exploring subsurface geology mainly involve drilling *boreholes*, which can be quite expensive. Thus, the availability of groundwater in a specific area is usually determined by combining many sources of information, including local knowledge and the performance of nearby wells. Selecting the location and depth of a well is sometimes an art as much as it is a science for groundwater hydrologists.

A well is usually installed by boring into the subsurface with a drill rig. The driller takes detailed records of the excavated soil and rock (called *cuttings*) that can be compared to assumptions about the geology that were made during the well design. Once a borehole is excavated to the proper depth, the well can be installed. Steel or plastic pipe, called CASING, is placed into the hole to provide support so that loose soil and rock can't fall into the well. A SCREEN is attached to the casing at the depths where water will be withdrawn. The screen allows groundwater to flow into the casing while keeping larger soil and rock particles out of the well where they could contaminate the water or cause additional wear on pumps.

Once the casing and screen are installed, the *annular space* (the area between the excavated borehole and casing) must be filled. Where the well is screened, this space is usually filled with gravel or coarse sand called GRAVEL PACK. This material acts as a filter to keep fine particles of the AQUIFER FORMATION from entering the well through the screen. The space along unscreened casing is usually filled with BENTONITE CLAY, which swells to create an impermeable seal so that shallower groundwater (which may be lower quality) can't travel along the annular space into the screens. Finally, the uppermost section of the annular space is permanently sealed, again using bentonite clay or sometimes using CEMENT GROUT. This seal ensures that contaminants on the surface cannot find their way into the well. In a worst-case scenario, pollutants could enter a well and flow into the aquifer, contaminating it for other users, so most jurisdictions have strict rules for sealing wells at the surface. The casing is usually extended aboveground to create a WELL-HEAD with a CONCRETE PAD extending in all directions to prevent damage or infiltration to the well.

The process of drilling a well can smear a layer of clay or fine particles along the surface of the borehole, creating an obstruction to flow. After a well is installed, it is usually taken through a procedure called *well development* to establish a hydraulic connection with the aquifer. Development involves surging water or air into and out of the well to remove fine sediments along the contact between the gravel pack and aquifer formation.

A properly completed and developed well allows groundwater to flow easily and sediment-free from the aquifer into the casing. However, it still needs a way to deliver that water up to the surface. Shallow wells can use *jet pumps* that draw water up using suction like a straw. However, this method doesn't work for deeper wells. When you drink through a straw, you create a vacuum, allowing the pressure of the surrounding atmosphere to push your beverage upward. However, only so much atmosphere is available to balance the weight of a fluid in a suction pipe. If you could create a complete vacuum in a straw, the highest you could draw a drink of water is around 10 meters or 33 feet. Thus, deeper wells cannot use suction to bring water to the surface. Instead, the pump must be installed at the bottom of the well so that it can push water to the top.

High-capacity wells are usually equipped with *vertical turbine pumps*. An ELECTRIC MOTOR is mounted to the wellhead and connected to a vertical SHAFT running down through the center of a COLUMN PIPE. At the bottom, the shaft drives a series of IMPELLERS that force water from the well up through the column pipe into the DISCHARGE LINE. Vertical turbine pumps are easy to service since the motor is accessible at the surface. However, they are noisy and require precise alignment of the well for its entire length. The popular alternative involves placing the motor at the bottom of the well with the impellers in a sealed assembly called a *submersible pump*. Submersible pumps are quieter because the moving components are deep below the ground, but they usually have a lower capacity since they use smaller motors to fit within the casing of a well.

If a pipe breaks or freezes, it may allow contaminated water to travel from the surface into a well, polluting the water within (and potentially even the surrounding aquifer). This is true not only for wells but also for water distribution systems. If a potable water supply loses pressure due to a main break or loss of power to pumps, hazardous pollutants can be drawn into the system. Backflow prevention devices are installed at wells and other locations in a water supply network where contaminants are present, such as irrigation systems and fire sprinklers. Many devices use two *check valves* in series to ensure water can only flow in one direction, even if one valve malfunctions. They are often combined with shutoff valves and ports to allow for regular testing of the mechanical components.

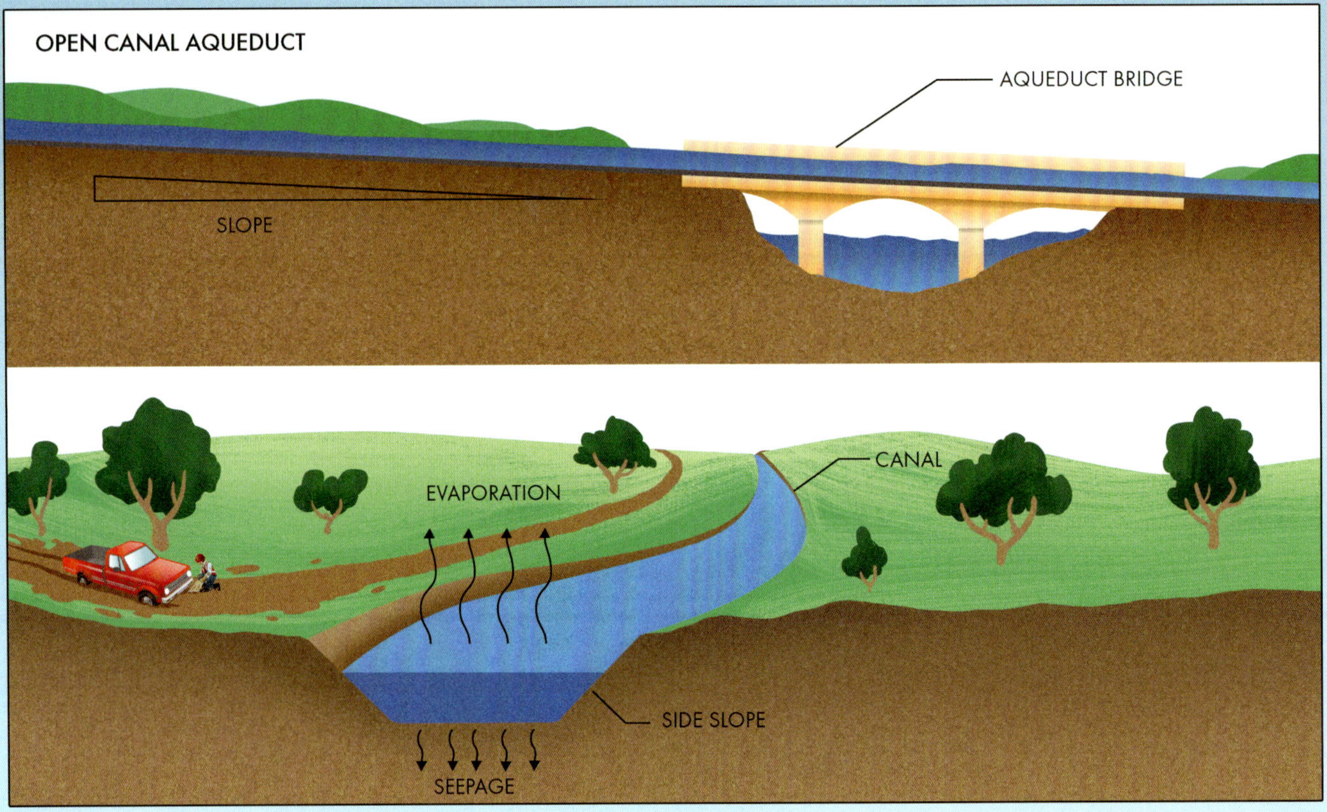

OPEN CANAL AQUEDUCT

AQUEDUCT BRIDGE

SLOPE

EVAPORATION

CANAL

SIDE SLOPE

SEEPAGE

UNDERGROUND AQUEDUCT

TUNNEL

SHAFT

TUNNEL LINING

SLOPE

INVERTED SIPHON

PRESSURIZED PIPELINE

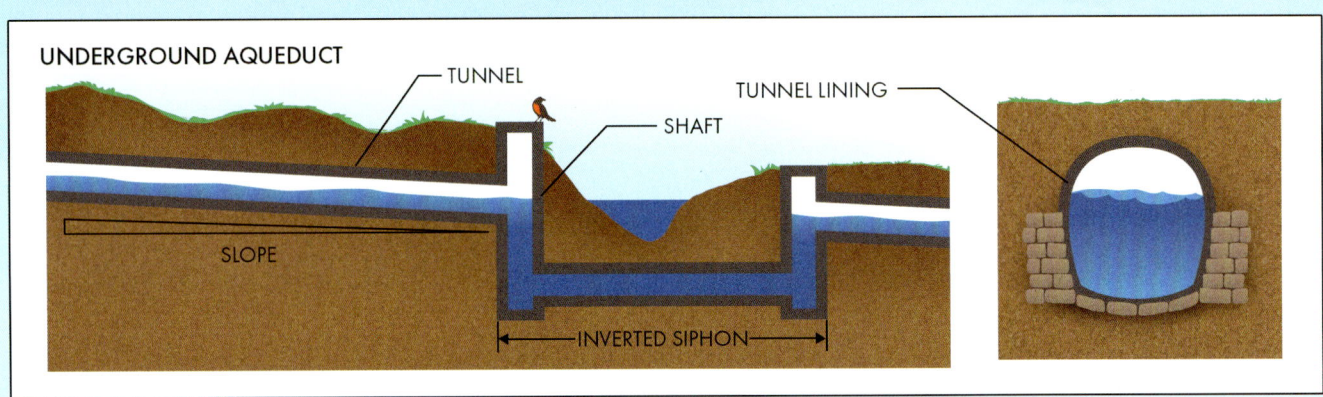

SURGE TANK

BOOSTER PUMP

GROUT

BELL

SPIGOT

GASKET

TRENCH

BACKFILL

COATING

PIPELINE

LINING

BEDDING

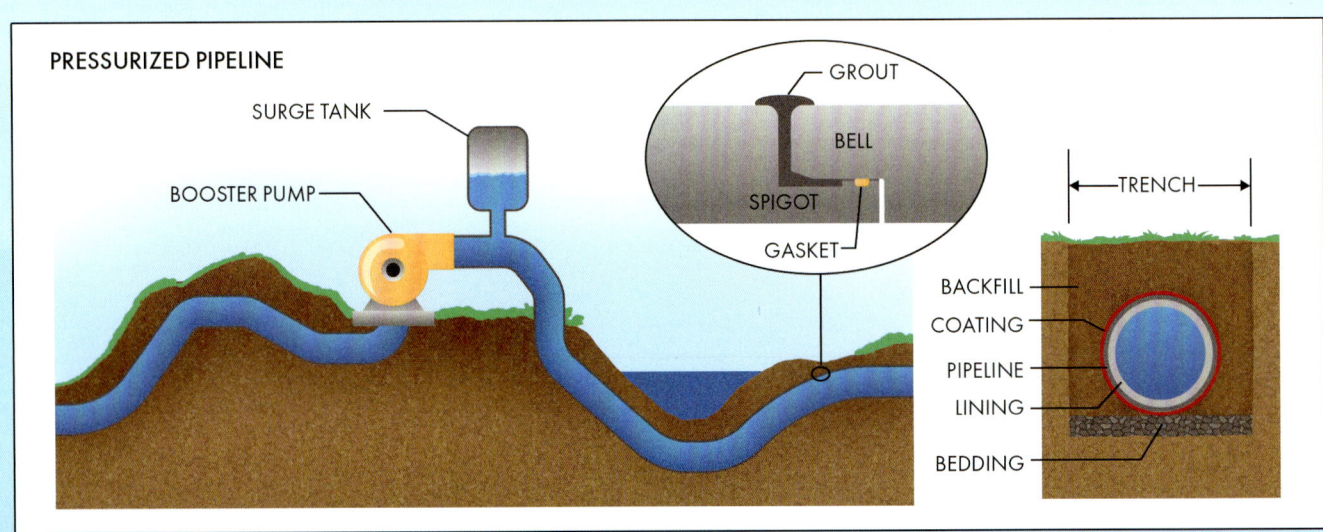

Transmission Pipelines and Aqueducts

Ideally, water resources are located near the place they're needed. Unfortunately, many populated areas don't receive ample precipitation throughout the year. As a result, some of the most impressive infrastructure projects in the world have the simple job of delivering *raw water* from a source to a populated region where it can be distributed to users. The ancient Romans were famous for their aqueducts, which traversed many miles to carry fresh water into cities, even crossing rivers using elaborate stone bridges. However, the BRIDGES were only a tiny part of each aqueduct system that often included miles of pipes, canals, and tunnels. Modern engineers use many of the same tools as the ancient Romans to move water to where it's needed.

Terminology can vary, but the word *aqueduct* generally describes any human-made structure meant to deliver water across a long distance. Perhaps the most straightforward technique for moving water is within an OPEN CANAL. If the source is high enough in elevation above its destination, excavating a channel is a guaranteed way to compel water to flow since gravity does all the work. Many aqueducts have very gradual SLOPES such that the incline is practically imperceptible to the eye. However, the amount of fluid that can flow by gravity is related to the channel's size and slope, so a steeper canal can be smaller (and thus cheaper to build) while moving the same volume of water as a larger, more gradual channel.

The flow volume isn't the only design consideration for an open channel, though. The flow velocity must be fast enough to minimize settling of silt on the canal floor but slow enough to avoid scouring and erosion. The channel also must be wide enough to carry sufficient flow without being so shallow as to accelerate the EVAPORATION of water into the air or SEEPAGE into the soil below. Engineers balance these factors when selecting the route a canal will take and the shape of the channel along the way. For example, many aqueducts run parallel to rivers where the topography naturally descends over long distances, and most canals use a trapezoidal cross section with SLOPED SIDES that are less likely to collapse. In addition, many canals include a concrete lining to mitigate seepage losses and scour.

Open canals are usually less expensive than other options, but they are subject to several disadvantages, including water loss through evaporation and seepage, freezing conditions that can block the flow, and their vulnerability to pollution. Canals also have environmental impacts because they divide the landscape just like a road or highway. Finally, channels can flow only downhill, limiting their practicality in hilly terrain. In many cases, moving an aqueduct underground in a TUNNEL or pipeline makes sense.

When not pressurized, an UNDERGROUND AQUEDUCT works exactly like a canal on the surface, flowing by gravity with a free surface at the top. The water is protected from pollution, evaporation, and seepage from the TUNNEL LINING or pipe. Underground aqueducts must have a consistent SLOPE for gravity flow, but that is easier to accomplish when the watercourse isn't constrained to the Earth's surface. Underground tunnels also reduce environmental issues by minimizing impacts at the surface. They can even travel under rivers using vertical SHAFTS, which help form INVERTED SIPHONS, eliminating the need for bridges.

When the source water is lower in elevation than its destination or the terrain along the way undulates too much for gravity flow, a PRESSURIZED PIPELINE may be the only feasible way for an aqueduct to function. A pumping station at the intake (as discussed in the previous section) forces water into the pipeline, allowing it to flow against gravity. These pipes are often installed in TRENCHES just far enough below the surface to protect against damage and freezing. The pipe is set atop a layer of BEDDING that acts like a mattress to distribute loads along the line.

The selection of the pipe material is a critical part of a pipeline design. The pipe must be strong enough to withstand both the internal water pressure and external forces from the BACKFILL and surface loads. Pipes must also resist corrosion from the water they carry inside and the soil along the outside. A pipe can be made from various materials, including steel, plastic, fiberglass, and concrete, and all materials have advantages depending on the situation. Larger pipelines often use protective outer COATINGS and inner LININGS to extend the lifespan of the line.

Unlike plumbing pipe that uses glue or threaded connections, most large diameter pipelines are either welded at each joint or use a bell and spigot design. When the SPIGOT of one pipe section slides into the BELL of another, it compresses a rubber GASKET, creating a watertight seal. A band of GROUT is sometimes installed around each joint to protect the gasket and any exposed steel from damage and corrosion.

Selecting the pipe size is another critical decision in the design of a pipeline. Smaller pipes are less expensive, but they require water to move faster to achieve the equivalent flow of a larger pipe. Water loses energy through friction, and these losses increase with velocity, so the money saved by installing a smaller pipe might be lost over time from the increased cost of pumping. For long pipelines, these frictional losses might be so significant to require BOOSTER PUMPS along the way that maintain pressure in the system. As pipes age, their inner surfaces get rougher, so engineers must consider the friction and pumping costs through the entire lifetime of the pipeline.

The mass of fluid in a long pipeline can be enormous, sometimes greater than a fully loaded freight train. When all that water is moving through a pipe, it has quite a bit of momentum. However, even though it's a fluid, water isn't very compressible, so closing a valve or stopping a pump gives that momentum nowhere to go. Instead, it creates a spike in pressure that travels as a shockwave through the pipe, an effect called *water hammer*. These shockwaves can be a problem in residential homes when taps close too quickly, resulting in lines knocking in the walls. However, in large pipelines that can contain immense volumes of fluid, closing a valve quickly can be the equivalent of slamming a freight train into a concrete wall. To avoid spikes in pressure that could damage equipment or rupture pipes, engineers specify slow-closing valves and pumps that start and stop gradually. In situations where operators need rapid control over the flow, SURGE TANKS can be installed to absorb the extreme pressure spikes and minimize the damaging effects of water hammer.

Although pipelines are meant to convey water, engineers also must consider what happens if air gets into the lines. Pipes are sealed systems, but air can still enter by being dissolved into water, introduced by pumps, or from the initial filling of the pipe. When these air bubbles coalesce in a high spot, they take up space and create a constriction to flow. In the worst-case scenario, air pockets can completely block a pipeline (an effect called *air lock*). Many pipelines are equipped with *air release valves* that can automatically vent bubbles from high spots in the pipe while keeping the water inside. You might see them protruding above the ground surface if you look carefully.

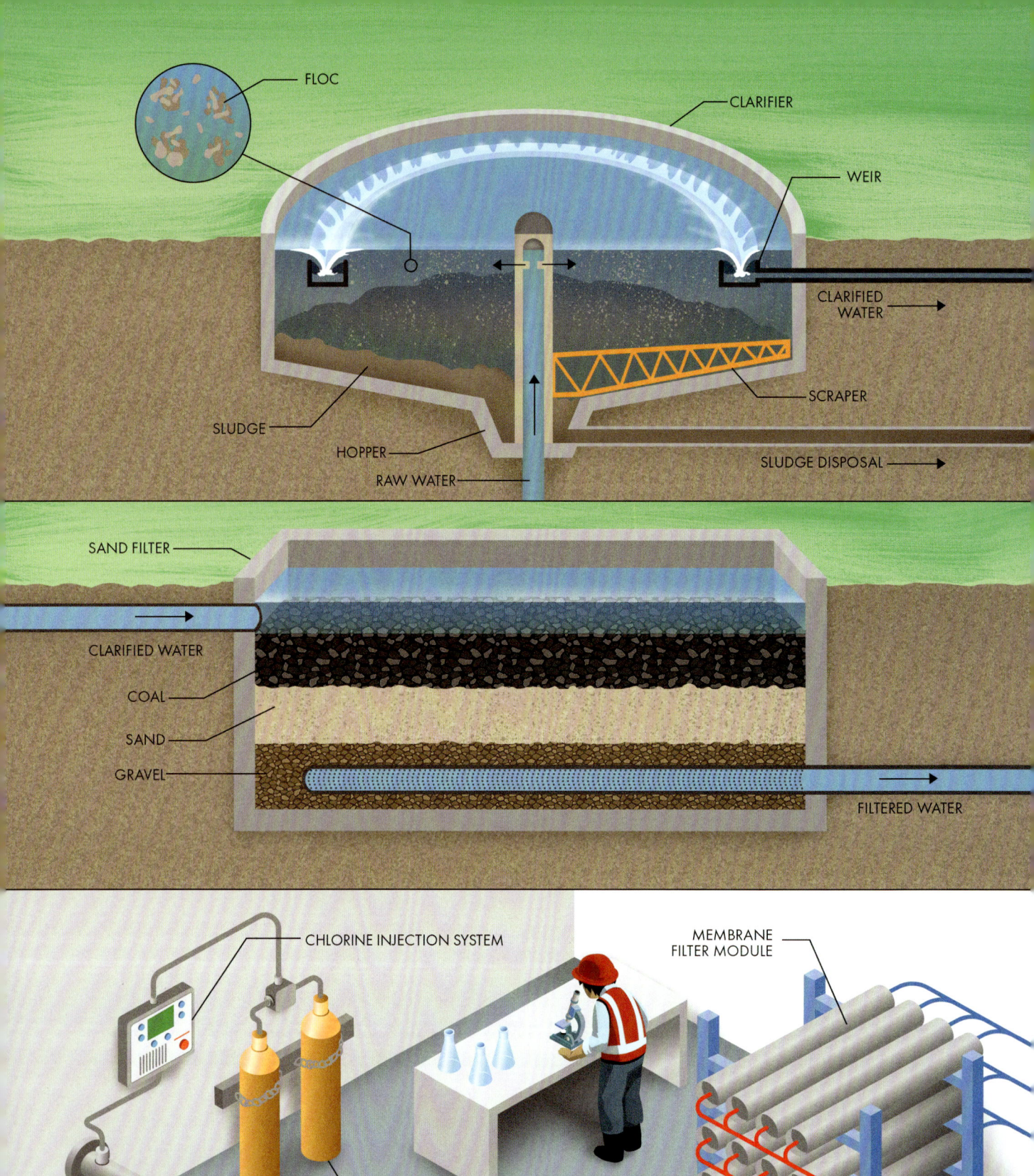

FLOC

CLARIFIER

WEIR

CLARIFIED WATER

SCRAPER

SLUDGE

HOPPER

RAW WATER

SLUDGE DISPOSAL

SAND FILTER

CLARIFIED WATER

COAL

SAND

GRAVEL

FILTERED WATER

CHLORINE INJECTION SYSTEM

MEMBRANE FILTER MODULE

CHLORINE CYLINDER

DISINFECTED WATER

Water Treatment Plants

Most sources of RAW WATER are subject to contamination by bacteria, sediments, and other substances that can be dangerous to human health. In addition, organic particles can negatively affect the water's taste and smell. Before water can be distributed to homes and businesses where it may be used for drinking or cooking, it usually must first go through a purification process at a water treatment plant to be made *potable*. A wide variety of techniques are used to purify water and ensure that it's safe for human consumption. Most water treatment plants are custom-designed for a specific water source and the potential contaminants that threaten it. For example, groundwater often requires less treatment than surface water sources because it is less vulnerable to pollution. Not every treatment plant uses the same set of processes, and not all parts of a water treatment plant are visible to an outside observer. However, understanding the basic steps of purifying water at a municipal scale provides insight and context into every other element of a city's water distribution system.

Groundwater and surface water both contain suspended particles of various materials. These solid particles give the water a murky appearance (called *turbidity*) and can harbor dangerous microorganisms. The first step in most treatment plants is to remove these suspended particles from the water in a process called *sedimentation*, which is often accomplished in a series of three stages. First, a chemical *coagulant* is vigorously mixed into the water. Coagulants neutralize the electric charges that cause suspended particles to repel each other, allowing them to stick together. Next, a chemical *flocculant* is added to the water, which bonds suspended particles together into groups called FLOCS. Flocculant is added to the raw water slowly to avoid breaking up the flocs.

As the flocs of suspended particles grow, they eventually become heavy enough to settle out (the third and final step of sedimentation). Raw water is pumped to a basin where it can sit nearly motionless while the flocs fall to the bottom. This basin can be as simple as a rectangular concrete box that's regularly drained and cleaned, but many water treatment plants use tanks called CLARIFIERS that include mechanisms to collect the solids as they settle at the bottom automatically. These circular basins are a recognizable component of many water treatment plants. The raw water flows up through the center of the clarifier and slowly makes its way toward the outer perimeter, dropping particles that form a layer of SLUDGE at the bottom. The clarified water passes over a WEIR so that only a thin layer farthest from the sludge can exit the basin. A SCRAPER pushes the sludge down the sloped bottom of the clarifier into a HOPPER where it can be collected for disposal.

Sedimentation removes most suspended solids, but it can't completely clean the water of tiny particles, viruses, and bacteria. Most water treatment plants follow sedimentation with a process of *filtration*, which involves forcing water to pass through porous media. The filters in municipal treatment plants commonly consist of layers of SAND, COAL, or other granular materials. The water flows by gravity or under pressure from pumps through the filter media while undesirable particles within the water are left behind. A layer of GRAVEL prevents any of the media from washing out with the filtered water. Over time, solids accumulate within the filter media, reducing its efficiency. Filters are *backwashed* by sending water through the opposite direction to clean the media. The backwash water is sent back to the treatment plant inlet to be reprocessed.

Some modern treatment plants have abandoned traditional sand filters for *membranes*, which consist of thin sheets of semipermeable material. Pressurized water is forced through the tiny pores in the membrane, leaving any unwanted particles behind. Water treatment plants using membrane filtration usually have a rack of tubular FILTER MODULES, allowing quick replacement of individual units if they clog or malfunction. Membrane filters can remove even the tiniest contaminants (including even viruses), so they are sometimes preferred over using multiple

separate treatment processes to create potable water.

The final step of a typical water treatment plant is DISINFECTION, where any remaining parasites, bacteria, and viruses are killed. There are several methods for deactivating microorganisms to make water potable, but the primary tool used in most cities involves adding a disinfectant chemical to the water (usually either chlorine or chloramine). These chemicals are safe for human consumption at low concentrations while still killing microorganisms that can make us sick. Many treatment plants use chlorine gas stored in metal tanks called CYLINDERS. An INJECTION SYSTEM carefully feeds the gas at a predetermined rate where it dissolves into the water and kills disease-causing pathogens.

An essential benefit of chemical disinfection is that it continues to work as water travels through miles of pipes from the treatment plant to individual customers within the distribution system. But before potable water leaves the treatment plant, it must first be tested to make sure it meets government standards for quality. Many different potential contaminants can be hazardous to human health, and source water chemistry can change over time (especially between seasons), so treatment plants must constantly verify that water at the outlet is clean and safe.

Chemical disinfectants, like chlorine, are usually added to water at the treatment plant. However, water quality standards require that some disinfectant remains in the water to the farthest reaches of the distribution system. This ensures that dangerous organisms cannot survive at any point along the way. The chlorine that stays in the water is called the *residual*, and it is a critical indicator that a water treatment and distribution process is operating effectively. Chlorine decays over time as it travels through pipes and tanks, but introducing disinfectant at a treatment plant offers only a single opportunity to provide enough residual at all points in a distribution system. Often, the pipes near the plant have too much chlorine and remote parts of the system have too little.

Many cities use booster chlorination stations in strategic locations to provide a more uniform distribution of disinfectant. Some can even automatically analyze the chlorine residual and adjust the booster dose accordingly. These booster stations may be located in small separate buildings or adjacent to other parts of a water distribution system (such as water towers or tanks). A warning sign about chlorine may be the only hint of what's inside.

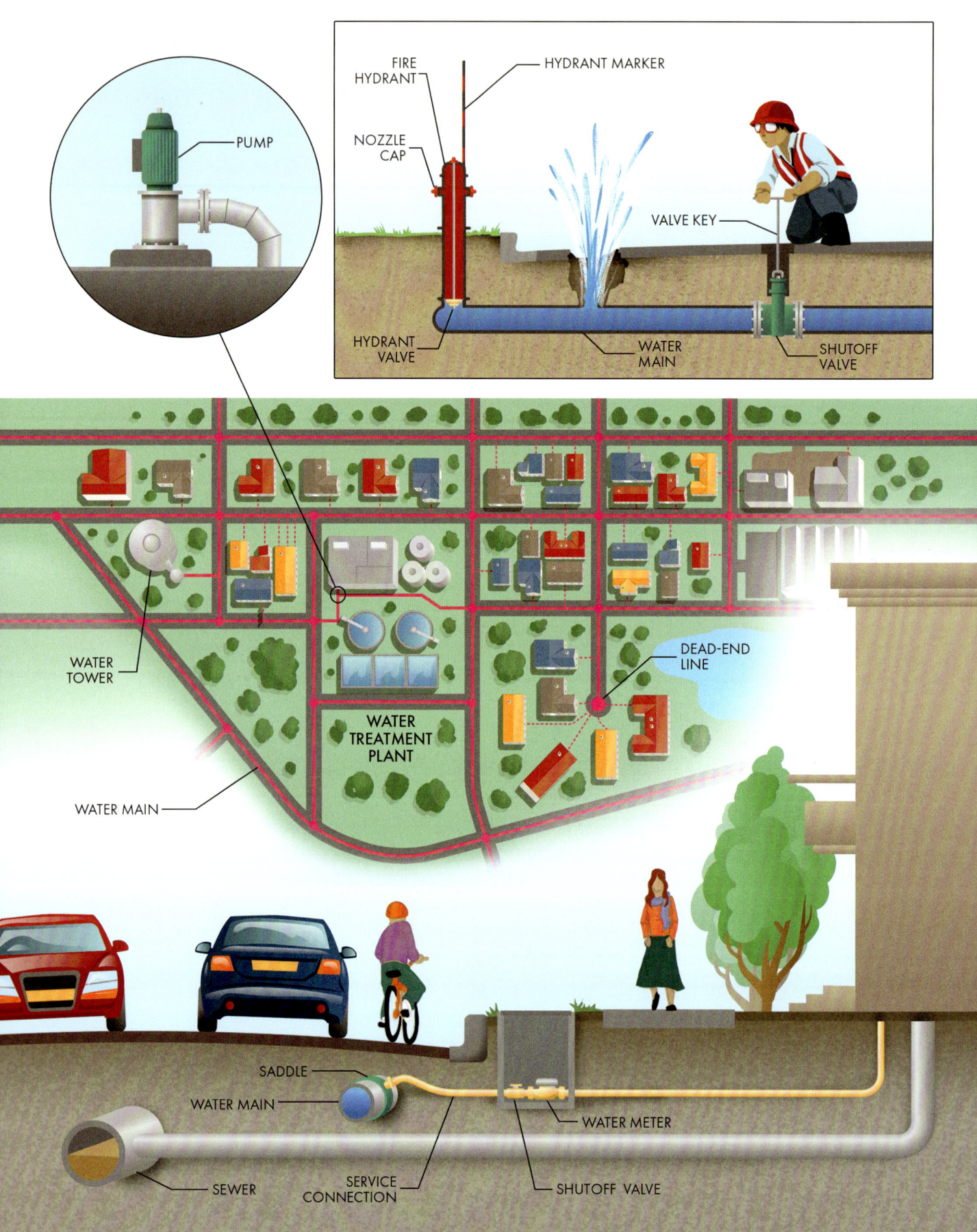

PUMP

FIRE HYDRANT

HYDRANT MARKER

NOZZLE CAP

VALVE KEY

HYDRANT VALVE

WATER MAIN

SHUTOFF VALVE

WATER TOWER

DEAD-END LINE

WATER TREATMENT PLANT

WATER MAIN

SADDLE

WATER MAIN

WATER METER

SEWER

SERVICE CONNECTION

SHUTOFF VALVE

Water Distribution Systems

Once water has been collected from a source, transported to a population center, and purified from contaminants, it must be delivered to the customers within a utility's service area. Potable water is transported from wells or treatment plants, sometimes across many miles, to each home and business. A municipal *water distribution system* consists of all the interconnected pipes, valves, and other elements that combine to carry clean water used for drinking, cleaning, cooking, watering plants, and a wide variety of commercial and industrial processes. The distribution system also provides a secondary benefit as a pressurized water supply for fighting fires to minimize their chance of spreading to adjacent structures. Unlike raw water infrastructure consisting of large, singular facilities, a water distribution system is necessarily spread out across an urban area. Many challenges are associated with constructing and maintaining such a sprawling system so critical to human health.

The first step in a water distribution system is usually the PUMPS. Like those located at the raw water intakes described previously, pumps do the job of pressurizing pipes within the system, usually to between two and six times normal atmospheric pressure, which is why they are often called *high-service pumps*. (Some of this pressure gets stored in tanks or WATER TOWERS, as discussed in the next section.) Pumping stations are usually located within a WATER TREATMENT PLANT to

send fresh water on its way. The pressure that pumps provide not only induces the potable water to flow to its destination but also ensures that contaminants can't enter the distribution system through open joints or small holes in the pipes. If leaks develop, water will flow out of the pressurized system rather than allow impurities or pollutants to enter. High-service pumps used in water distribution systems consume significant amounts of electricity, so they often require robust connections to the electrical grid and backup generators for potential outages. Energy is often one of the highest ongoing costs for a water utility. Conserving water reduces waste of the water itself and reduces the significant amount of energy required to collect, clean, and deliver it.

From the pumps, clean water enters a series of pipes called WATER MAINS, the circulatory system for potable water within a city. Water mains are usually installed belowground to protect them from damage and, more important, from freezing. Most mains are connected in a grid or loop pattern, often following the paths of city streets. Many jurisdictions require that water mains be horizontally separated from underground SEWERS, so when these lines run parallel, they are often located on opposite sides of a street.

Installing water mains in a *gridded* fashion requires additional pipes and joints. However, in a gridded system, water can take multiple paths to any given location, increasing the reliability of service and allowing water mains to be repaired without affecting the rest of the network. Gridded water mains also help avoid stagnation. When pipes have DEAD-ENDS, water flows only when users along each particular line turn on the tap. If clean water sits in a pipe for an extended duration, the disinfectant can decay, deteriorating the water quality. In gridded systems, the water in the pipes continuously circulates to meet demands wherever they occur.

Individual customers get their water from the main lines through SERVICE CON-NECTIONS. A SADDLE is used to create a tap point to the main line. Piping usually runs from the saddle to a WATER METER that measures the volume of water used, allowing the utility to charge each customer based on usage. Metering each service connection encourages water conservation and helps utilities identify leaks in the distribution system.

Water mains do break on occasion, usually due to shifting ground, freezing, or simple deterioration due to age. When this happens, the pipe must be excavated and repaired. Although repairing a line while it's discharging a geyser of water is possible, it's usually a difficult ordeal. Isolating the main from the rest of the system before starting repairs is much easier. SHUTOFF VALVES are typically located at intersections in water mains to allow parts of the network to

be disconnected so crews can fix a broken pipe. Valves are installed in belowground enclosures with small steel lids. Most pipe intersections have one line without a valve to save installation and maintenance costs. If the unvalved pipe must be isolated, all the other valves at the intersection are shut. Crews use a VALVE KEY to open or close each valve, as needed. Similarly, one or more shutoff valves are also included at each service connection so that individual homes or businesses can be isolated for plumbing repairs or during an emergency.

Although clean water is essential for basic human needs, cities also need water readily available for fighting fires. Some of the worst disasters in history occurred when fire raced through a populated area with insufficient means to prevent the spread. Cities are dotted with FIRE HYDRANTS that provide connection points to pressurized water mains to help extinguish fires. Most places in the United States use dry-barrel hydrants, which have their VALVES located belowground, protecting them from damage due to errant vehicles and reducing the chance of water freezing within the exposed hydrant. In some places, the color of the hydrant NOZ-ZLE CAPS indicates the maximum flow rate available for firefighting. In colder areas, hydrants may include MARKERS that extend above the snowfall to make them easier to locate in the winter.

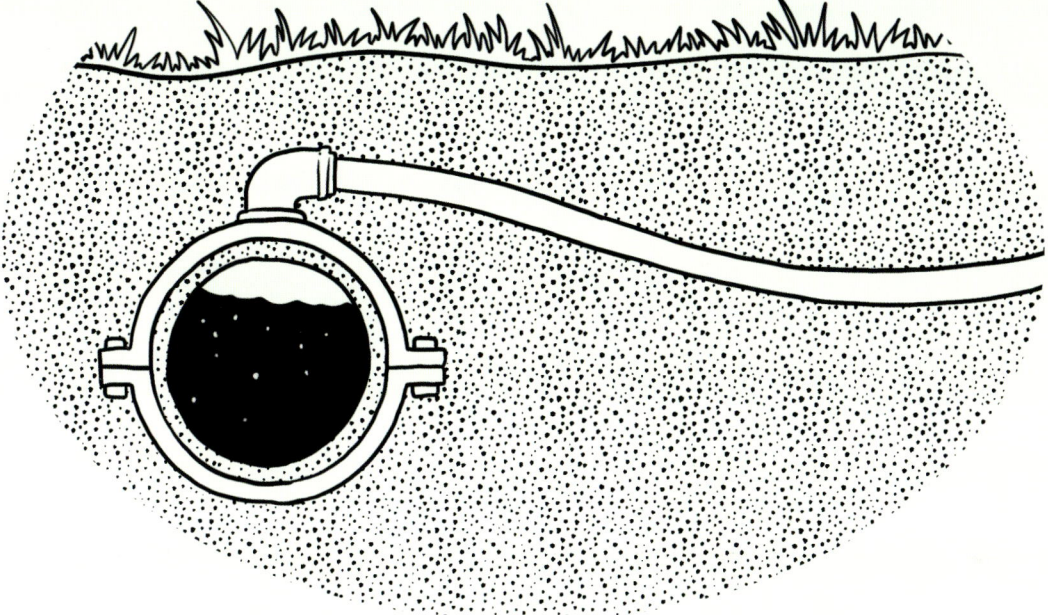

Until the early 20th century, it was common to use pipes made from lead to connect houses and businesses to underground water mains, and some cities continued to allow lead service lines into the 1980s. Lead is a durable metal while still being flexible enough to make pipes easy to bend. However, even at low concentrations, exposure to lead is dangerous to human health and can cause neurological effects, especially in children. Lead can leach into water traveling through pipes, exposing humans to this hazardous contaminant. Most cities with large numbers of lead service lines are working to replace them permanently, often at a high cost. In addition, some cities introduce corrosion-inhibiting chemicals in the water to reduce the chance of leaching lead from the pipes before they can be replaced. If you're unsure if there is lead in your water, consider having it tested by a laboratory to reduce your chance of exposure to this dangerous heavy metal.

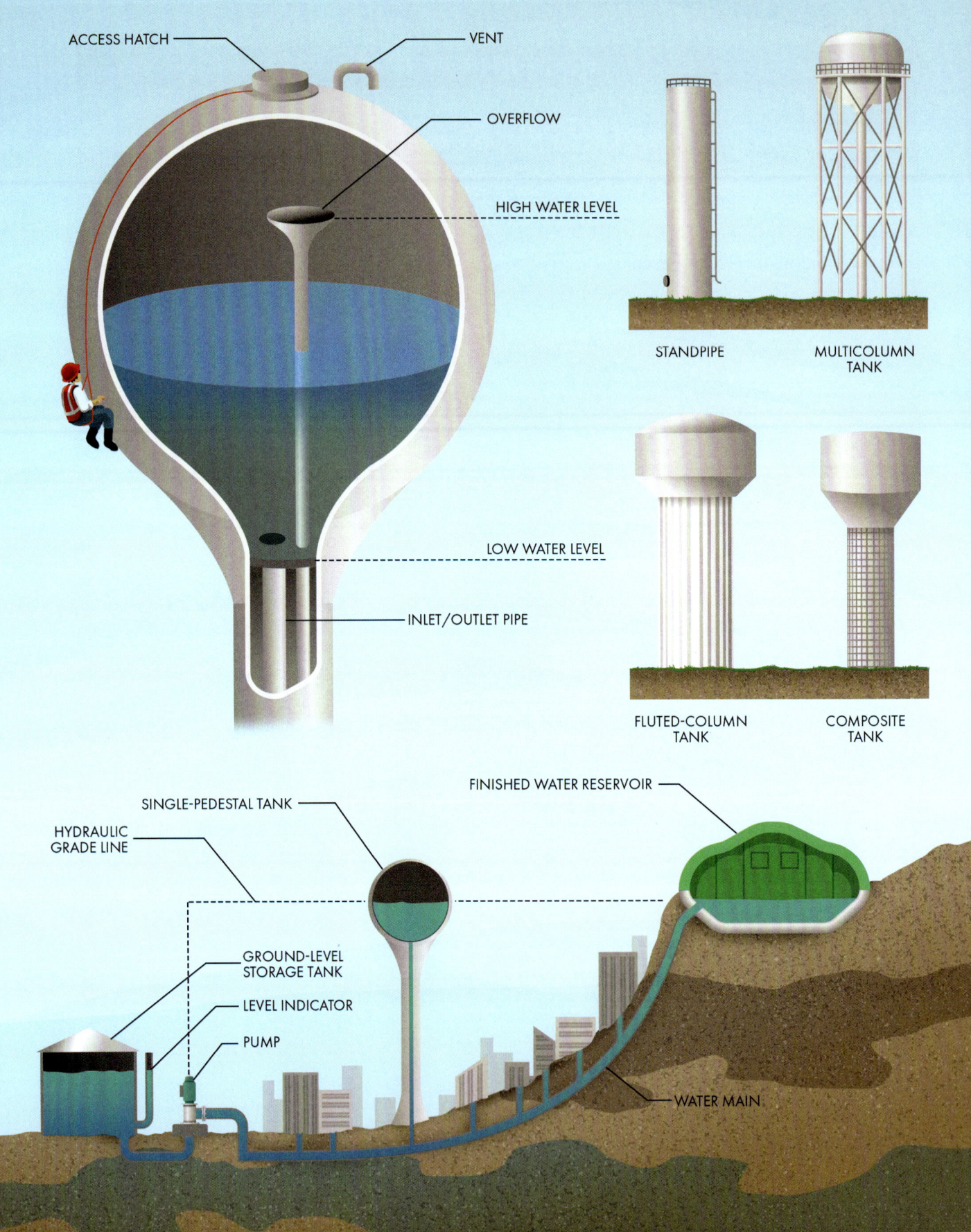

ACCESS HATCH

VENT

OVERFLOW

HIGH WATER LEVEL

LOW WATER LEVEL

INLET/OUTLET PIPE

STANDPIPE

MULTICOLUMN TANK

FLUTED-COLUMN TANK

COMPOSITE TANK

FINISHED WATER RESERVOIR

SINGLE-PEDESTAL TANK

HYDRAULIC GRADE LINE

GROUND-LEVEL STORAGE TANK

LEVEL INDICATOR

PUMP

WATER MAIN

Water Towers and Tanks

The demand for clean water varies significantly not only over the course of a year (due to seasonal changes in weather) but even over a single day. Water use in a city is usually highest in the mornings and evenings when people are taking showers, cooking, and watering lawns. In addition, some of the highest water demands in cities are created by fires, which occur randomly, day or night. Fires can burn out of control in dense urban areas, so most municipalities ensure their distribution systems have reserve capacity even on the days with the highest water demands. Engineers who design distribution systems must consider all the possible variability in flows when selecting the sizes for pumps, pipes, valves, and other equipment. One of the most important parts of a water distribution system (and often the most visible) is a solution to all this variability in potable water demand: storage.

Many steps involved in collecting, moving, purifying, and distributing water are most efficient when they happen at a steady rate. At treatment plants, chemical feeds and purification processes can't tolerate sudden changes. In addition, PUMPS used in water distribution systems often run at a single speed. Without somewhere to store water, operators would constantly need to ramp production up or down to meet changing demands. In addition, all the treatment facilities and pumps would need to be sized for peak water needs, even if they were used to their total capacity only once or twice per year, increasing their cost and complexity.

Tanks and reservoirs smooth out the peaks and valleys of water demand, allowing pumps and other infrastructure to be operated for average conditions. When usage is low (for example, at night), the treatment plant overproduces to fill the tanks. When use is high, the stored water can supplement the treatment plant to meet water demands.

Many types of storage structures are used in water distribution systems. GROUND-LEVEL TANKS often consist of large, circular steel or concrete enclosures. If you look closely, many tanks include a LEVEL INDICATOR on the outside to show the water level at a glance. Some cities use ground excavations to form ponds called FINISHED WATER RESERVOIRS that hold significant volumes at a relatively low cost. These ponds are often lined with plastic or concrete to prevent leakage and are covered to minimize the chance of contamination (although some uncovered reservoirs still exist today). Both ground-level tanks and reservoirs can often be seen at water treatment plants, where they are known as *clearwells*.

One disadvantage of storing water at ground level is that it is not pressurized and therefore must be pumped into a distribution system, fluctuating water demands. Tanks or reservoirs are often installed at the tops of hills or mountainsides above the area served by the system, allowing storage not only of the water but also the energy imparted to it by the pumps. *Elevated storage* smooths the demands on pumps, allowing them to run consistently instead of cycling on and off to meet changing water demands throughout the day. In some areas where electricity costs vary, pumps can run at night when power is cheap to fill up tanks, and be left off during peak hours when electricity is more expensive. Elevated storage is also beneficial during an electrical outage or emergency, keeping pipes pressurized and water flowing even when the pumps or treatment plant is out of service.

Unfortunately, not all cities have hills or mountains on which water tanks can be built. Smaller distribution systems often use tall, narrow tanks called STANDPIPES for potable water storage. The water at the top of the tank serves as elevated storage just as if it were sitting on a hilltop. The water at the bottom of the tank serves as an emergency reserve that can be pumped into the distribution system if needed. Large cities often use elevated storage tanks, also known as *water towers*, that have their entire storage volume well above the minimum system pressure.

Choosing the height of a storage tank is an important decision. Distribution systems must be maintained within a window of acceptable water pressures. Too low, and you risk potential contamination. Too high, and you risk damage to pipelines and equipment.

The pressure in a body of water is related to the depth below the surface. You can imagine a water distribution system as a virtual ocean under which we all live. The water surface in elevated storage tanks represents the surface of the virtual ocean

(called the HYDRAULIC GRADE LINE by engineers). Customers at low elevations are at the bottom of the virtual ocean, where pressures are highest, and customers at high elevations will be near the surface of the virtual ocean where pressure is lowest. The ideal depth is usually around 30 to 60 meters (approximately 100 to 200 feet), which means most water towers are set so that their LOW and HIGH WATER LEVELS are within this range. Water stored less than 15 meters (around 50 feet) above the system may not create enough pressure to prevent potential contamination. Cities with large changes in elevation sometimes maintain independent water distribution networks at different pressures to keep customers within the ideal range.

A water tower is as simple as a tank connected to a WATER MAIN. When water demands fall below the pump capacity, the pressure in the system goes up, forcing water into the tank through the INLET/OUTLET PIPE. When demands rise above the pumping rate, the system pressure goes down, and water flows out of the tank through the same pipe to supplement water from the treatment plant. Other than water, there's not much inside. Most tanks feature an OVERFLOW to prevent them from overfilling. VENTS make sure that the air pressure in the tank doesn't change with the water level, potentially creating positive or negative pressure that could damage the structure. ACCESS HATCHES provide means for maintenance and inspection inside the tank.

Water towers come in a wide variety of designs. They're most often described either by the shape of the tank itself or the tower structure on which it sits. SINGLE-PEDESTAL TANKS and MULTICOLUMN TANKS are usually made entirely from welded steel. FLUTED-COLUMN TANKS are supported by corrugated steel with lots of room inside the tower for equipment storage and sometimes even offices. COMPOSITE TANKS sit upon concrete towers, saving the cost of regular painting required for steel columns to protect against corrosion. For cities that use elevated storage, these tanks are often a central part of operating the entire system. The water level in the tank is the primary indicator that the distribution system is pressurized to the right level and functioning as designed to deliver clean water to each individual customer.

KEEP AN EYE OUT

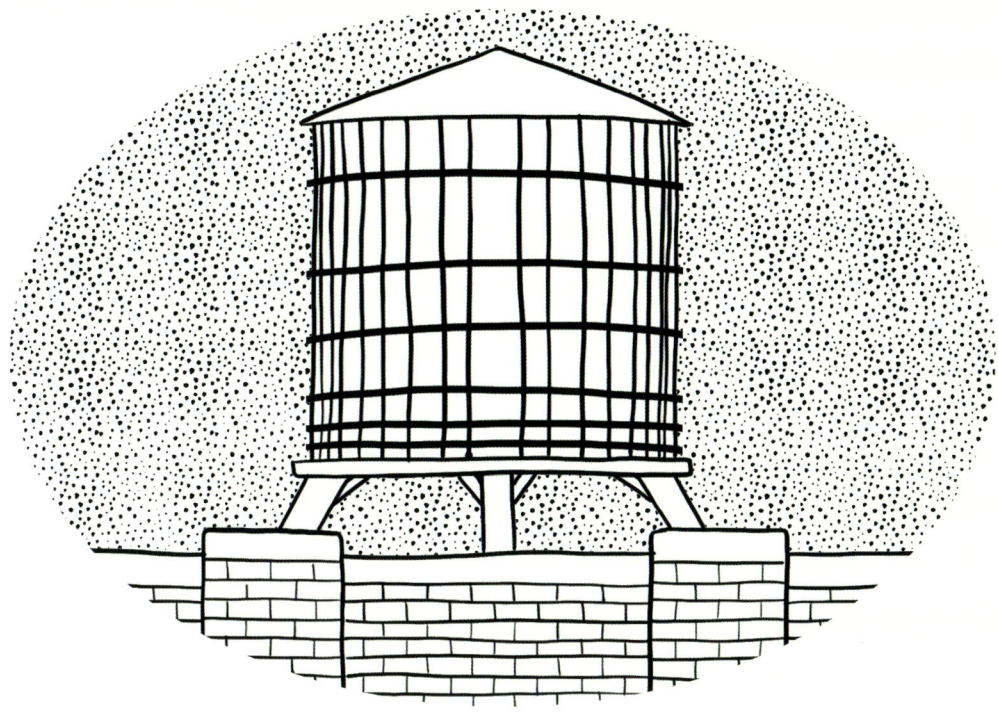

In large cities, it's not unusual for buildings to be so tall that the main water pressure cannot deliver water to the top. Most tall buildings have their own system of pumps and tanks to ensure that each floor has adequate water pressure. Some cities require buildings to have a rooftop tank and pump, effectively spreading out the elevated storage across an urban area (rather than having centralized large towers). These rooftop tanks are often constructed from wood because it's inexpensive and provides insulation against freezing. Steel bands hold the wooden boards tight against the pressure of the water inside. The bands' spacing decreases toward the bottom of the tank, where the pressure is highest.

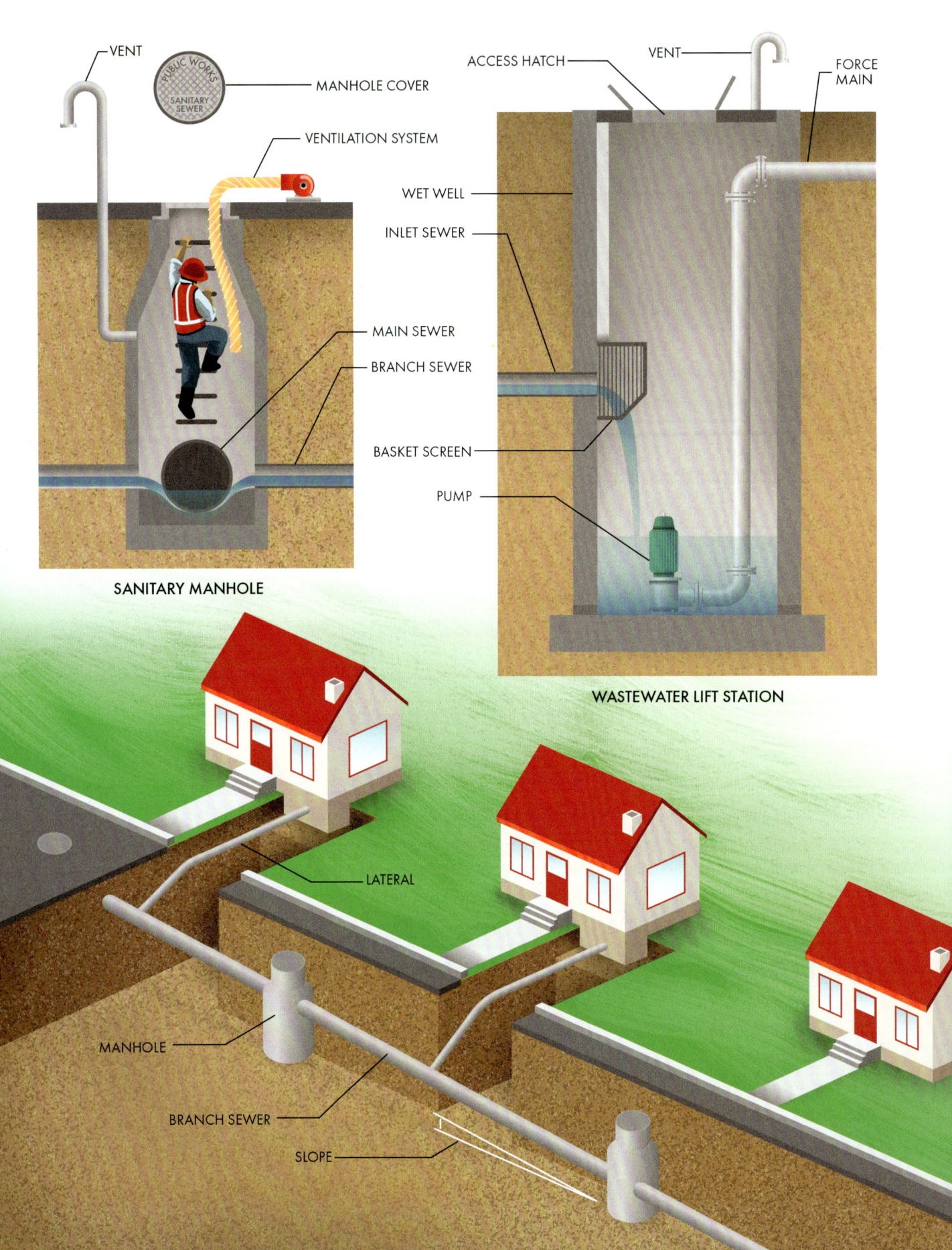

VENT

MANHOLE COVER

PUBLIC WORKS
SANITARY
SEWER

VENTILATION SYSTEM

MAIN SEWER

BRANCH SEWER

SANITARY MANHOLE

ACCESS HATCH

VENT

FORCE MAIN

WET WELL

INLET SEWER

BASKET SCREEN

PUMP

WASTEWATER LIFT STATION

LATERAL

MANHOLE

BRANCH SEWER

SLOPE

Sanitary Sewers and Lift Stations

Humans are kind of gross. We collectively create a constant stream of waste that threatens city dwellers with deadly diseases unless it's safely carried away. A lot of technical challenges are involved with getting so much poop from point A to point B, and the fact that we do it primarily out of sight and mind, I think, is cause for celebration. *Sanitary sewers* convert that figurative stream into a literal one that flows belowground away from public view (and hopefully public smell). The original sewers were simply rivers and creeks into which waste was thrown to be carried away downstream. That method of managing sewage had some obvious limitations, including that it contaminated sources of water that were often used for drinking. Modern sewers are almost always installed as pipes below the ground to keep waste streams separate from drinking water sources, but they still function much like the watercourses on the surface.

Sewers rely on gravity to do the work of collecting and transporting waste, flowing downward, converging, and concentrating into larger and larger streams. Sanitary sewer networks are *dendritic* with small pipes at individual buildings concentrating into larger and larger lines until all the wastewater converges at a single treatment plant. Pipes that service individual buildings are usually called LATERALS, and those servicing particular streets are BRANCHES. Larger pipes that collect wastewater from multiple branches are called MAIN SEWERS or *trunk sewers*. The most significant lines and the farthest downstream in the system are usually called *interceptors*.

Sloping sewers to flow naturally downhill is convenient because we don't have to pay a gravity bill, and it doesn't get knocked out during a thunderstorm. However, relying only on gravity also constrains sewer design and construction. If the sewage flows too quickly, it can damage joints and erode the walls of the pipes. However, if the sewage flows too slowly, solids can settle out of suspension and build into clogs and constrictions. We can't adjust gravity up or down to maintain this balance of flow velocity, and we also don't have much control over the volume of wastewater (since people flush when they flush). The only factors engineers can control are the size of the sewer pipe and its SLOPE. Each sewer line is carefully sized and sloped according to the expected quantity of wastewater to keep it flowing steadily toward a treatment plant.

Any time the size or direction of a sewer changes and at intersections of pipes, a SANITARY MANHOLE is installed to provide access for maintenance and inspection. Manholes are usually made from vertical concrete enclosures that run up to the ground surface. Steps allow personnel to enter and exit. A heavy cast iron plate, called a COVER, keeps people and debris out of the sewers while allowing vehicles to drive over the top. Manholes also sometimes serve as vents to equalize the air pressure within the lines and prevent the buildup of toxic gases. When the top of a maintenance hole is vulnerable to flooding, cities often require that the cover be sealed and bolted shut to exclude rainwater from entering the pipes. In this case, VENTS sometimes extend above the required flood level to keep air pressure from building up, even during heavy storms. A temporary VENTILATION SYSTEM is used whenever someone enters a manhole to provide fresh air during repairs or maintenance.

Because sewers must always slope, they often end up well below the ground surface, especially toward their downstream ends, making construction costly and time-consuming. In some cases, it isn't feasible to chase the slope of a sewer farther and farther below the ground surface. One alternative is to install a pumping station that can lift raw sewage from its depths back up and closer to the surface. LIFT STATIONS can be small installations designed to handle a few apartment complexes or massive projects that pump significant portions of a city's wastewater flow. A typical wastewater lift station consists of a concrete chamber called a WET WELL. Sewage flows into the wet well by gravity via the INLET SEWER, filling it over time. Once the level reaches a prescribed depth, a PUMP turns on, pushing the wastewater into a pipe called a FORCE MAIN. This intermittent operation makes sure that sewage is always moving swiftly through the line so that solids don't settle out of suspension during off-peak hours. The sewage travels under pressure within the force main to an uphill manhole where it can once again continue its journey downward via gravity. Lift stations usually feature

multiple pumps so that they can continue operating if one fails. They often have backup generators so that the sewage can continue to flow even if grid power is lost.

We often think of sewage as its grossest constituents: human excrement. But sewage is a slurry of liquids and solids from a wide variety of sources. Lots of stuff ends up in our wastewater stream, including soil, soap, hair, food, wipes, grease, and trash. These things may make it down the toilet or sink drain and through the plumbing in your house with no problem. However, in the sewer system, they can conglomerate into large balls of debris (sometimes called *pigtails* or *fatbergs* by wastewater professionals). In addition, with many cities putting efforts into water conservation, the concentration of solids in wastewater is trending upward. Conventional pumps handle liquids just fine, but adding solids in the stream increases the challenge of lifting raw sewage. Pumps used in wastewater lift stations are designed for extra wear and tear, but no pump is entirely clog-proof.

One solution to the problem of clogging is to use a screen in the lift station wet well to prevent trash from reaching the pumps. Every so often, the garbage caught in the mesh must be removed from the wet well and hauled away to a landfill. Smaller lift stations often use a BASKET SCREEN on rails that can be manually lifted through an ACCESS HATCH at the surface. Larger pump stations may have automatic systems that remove solids from the screen into a dumpster. Another solution to debris in the wastewater stream is to grind it into smaller pieces. Some lift stations feature *grinders* that chew through debris so it can't clog the pumps, minimizing the need for staff to visit the station to perform repairs or remove trash. The solids remain in the wastewater stream to be removed farther down the line at a treatment plant (discussed in the next section).

Most sanitary systems are separated from storm drains, which carry away rainfall and snowmelt. However, precipitation can still make its way into the sewage system. *Inflow and infiltration* (often simplified as *I&I*) are the enemies of utility providers for one simple reason: precipitation finding its way into sewers can overwhelm the system's capacity during storms. I&I can lead to overflows that create exposure to raw sewage and environmental problems, so municipalities work hard to find and repair defects that allow rainwater to enter the sewers. Cities often perform regular inspections of sewer lines using video cameras that can travel through the pipes on remotely controlled vehicles. Another type of inspection involves introducing nontoxic smoke into sewers as a way to detect sources of I&I. The smoke can be seen escaping through openings and defects, allowing for visual identification of cracks, breaks, faulty manhole seals, and illicitly connected stormwater drains.

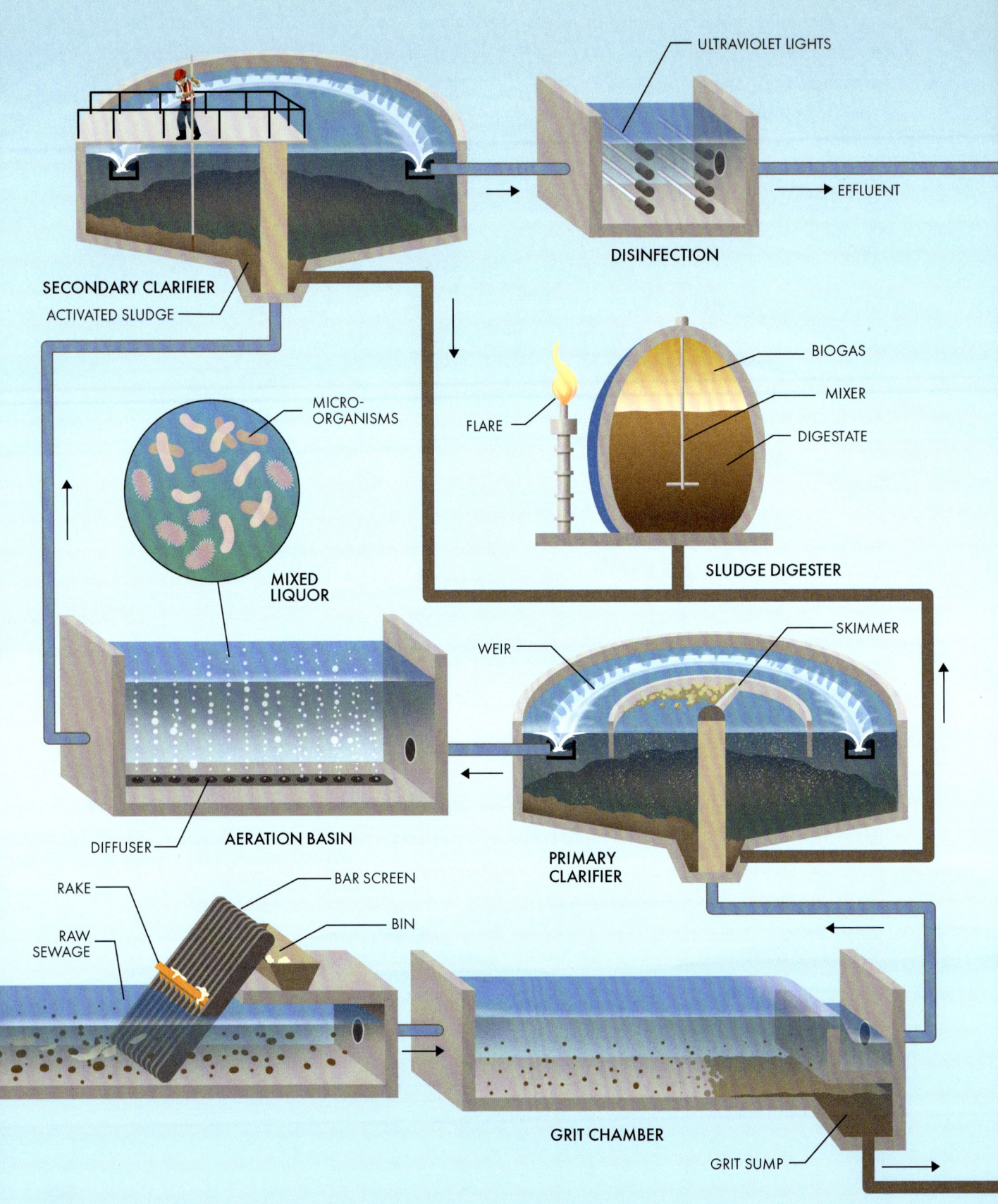

ULTRAVIOLET LIGHTS

EFFLUENT

DISINFECTION

SECONDARY CLARIFIER

ACTIVATED SLUDGE

BIOGAS

MIXER

DIGESTATE

FLARE

SLUDGE DIGESTER

MICRO-ORGANISMS

MIXED LIQUOR

WEIR

SKIMMER

DIFFUSER

AERATION BASIN

PRIMARY CLARIFIER

RAKE

BAR SCREEN

BIN

RAW SEWAGE

GRIT CHAMBER

GRIT SUMP

Wastewater Treatment Plants

Water gets along with nearly every substance on Earth, which is a significant part of why it does such an excellent job of carrying our wastes away from homes and businesses in sewers. Before modern environmental regulations, it was not uncommon for a city to discharge RAW SEWAGE into rivers to be carried away downstream. Now nearly all wastewater collection systems rely on some kind of treatment plant to reverse the dirtying process, removing contaminants from the water so it can be reused or released back into the environment. Technology continues to evolve, and a wide variety of processes are used in wastewater treatment around the world. This section discusses some of the most common treatment methods modern wastewater plants use. If you can cope with the smell, many municipal treatment plants are happy to offer tours to the public where you can see each process in action.

Wastewater treatment plants use many discrete steps in the process of cleaning sewage. Many of these steps are similar to those used in drinking water treatment plants (discussed previously). However, the standards are generally lower since processed water (called EFFLUENT) will not be used for human consumption but only must be safe enough to release into the environment. The initial steps in a water treatment plant, called *primary treatment*, involve the physical separation of contaminants that are suspended in the swift and turbulent flow. First, sewage flows through a BAR SCREEN that filters out large debris like sticks, rags, and any other large pieces of detritus that find their way into the sewers. Many technologies exist, including bar screens equipped with automated RAKES that scrape the filtered debris into a BIN to be discarded as solid waste.

Next, suspended particles are separated from the flow. Sand and soil found in wastewater are collectively known as *grit*. These materials can damage equipment in the plant, so they are usually removed in a separate process during primary treatment. Treatment plants use GRIT CHAMBERS, often configured as long narrow tanks, to slow down wastewater flow. Within these placid conditions, suspended sediments will settle to the bottom of the chamber while the grit-free wastewater continues to the outlet. Some grit chambers introduce air bubbles that help fling heavier particles to the edges of the tank. Others use a motorized agitator to create a vortex in the flow to accomplish a similar feat. A SUMP at the bottom of the chamber collects the settled grit to be pumped away for disposal.

The final step in primary treatment is also usually a gravity process. Wastewater leaving the grit chamber is still full of suspended solids, but they mainly consist of small, organic particles or floating oils and grease (collectively known as *scum*). Most treatment plants use PRIMARY CLARIFIERS to separate these remaining solids. The large circular tanks slow down wastewater flow even further, allowing tiny particles to sink gently while a SKIMMER collects the solids floating on the surface. The solids are sent for further processing, and the clarified wastewater exits over a WEIR into the secondary treatment process.

Where primary treatment physically separates contaminants from wastewater, *secondary treatment* does so using biological processes, replicating what Mother Nature would do naturally but in a much shorter time period. Most wastewater treatment plants take advantage of microorganisms that can digest organic matter in sewage. As they consume contaminants, these bacteria and protozoa clump together, leaving relatively clean water behind. The microorganism communities that thrive in oxygen-rich (*aerobic*) environments are different from those that live in oxygen-depleted (*anaerobic*) environments. These various colonies consume different nutrients from the water, so treatment plants often utilize both aerobic and anaerobic conditions to remove wastewater contaminants thoroughly. Aerobic conditions are created in AERATION BASINS where blowers generate a constant supply of air that passes through DIFFUSERS, creating tiny bubbles that mix and dissolve oxygen into the water.

Once the biological treatment has consumed most of the nutrients, the cleaned water with suspended clumps of MICRO-ORGANISMS (called MIXED LIQUOR) moves from the aeration chamber into a SECOND-ARY CLARIFIER. Here, the colonies of bacteria settle to the bottom, allowing only the clean effluent to be discharged. Depending on the regulatory requirements, many treatment plants have tertiary processes that target specific contaminants. In addition, most treatment plants perform a final DISINFECTION to kill any remaining pathogens in the water. Disinfection may be completed using dissolved chlorine, ozone gas, or intense ULTRAVIOLET LIGHTS, which deactivate viruses and harmful bacteria. The final effluent from the treatment plant is usually discharged into a natural stream, river, or ocean.

Some of the microorganisms that settle out in the secondary clarifier (known as ACTIVATED SLUDGE) are returned to the aeration chamber to seed the next colony. The rest of the sludge must be discarded. Some treatment plants send sludge directly to a landfill for disposal. However, sludge is an organic material and will decompose over time, releasing unwanted gases like methane into the environment. Rather than allow such decomposition to occur in a landfill, many treatment plants use DIGESTERS to process organic solids. Digesters convert sludge into BIOGAS, which can be used as a fuel for heating or electricity generation, and a solid material called DIGESTATE (or *biosolids*), which can be dried and landfilled or used as fertilizer. Digesters often have MIXERS to keep sludge blended, large domes to collect biogas as it is generated, and a FLARE that serves as a safety measure. If too much biogas is created than can be stored, operators will allow the gas to combust in the flare, converting the harmful constituents into safer gases to be released into the environment.

Raw sewage is 99.9 percent water, and water is a valuable resource to cities. In places where water is scarce, it can be cost-effective to treat municipal wastewater beyond what would typically be required so that it can be reused instead of discarded. A few places across the world use *direct potable reuse* (colloquially known as *toilet-to-tap*) where sewage is cleaned to drinking water–quality standards and reintroduced to the distribution system. However, most recycled water is not meant for human consumption. Plenty of uses do not require potable water, including industrial processes and the irrigation of golf courses, athletic fields, and parks. Many wastewater treatment plants are now considered water reclamation plants because, instead of discharging effluent to a stream or river, they pump it to customers that can use it, hopefully reducing demands on the potable water supply as a result. In many countries, purple pipes are used to distinguish non-potable water distribution systems, helping to prevent cross-connections. In addition, users of recycled water will often post signs warning the public that irrigation water is not safe to drink.

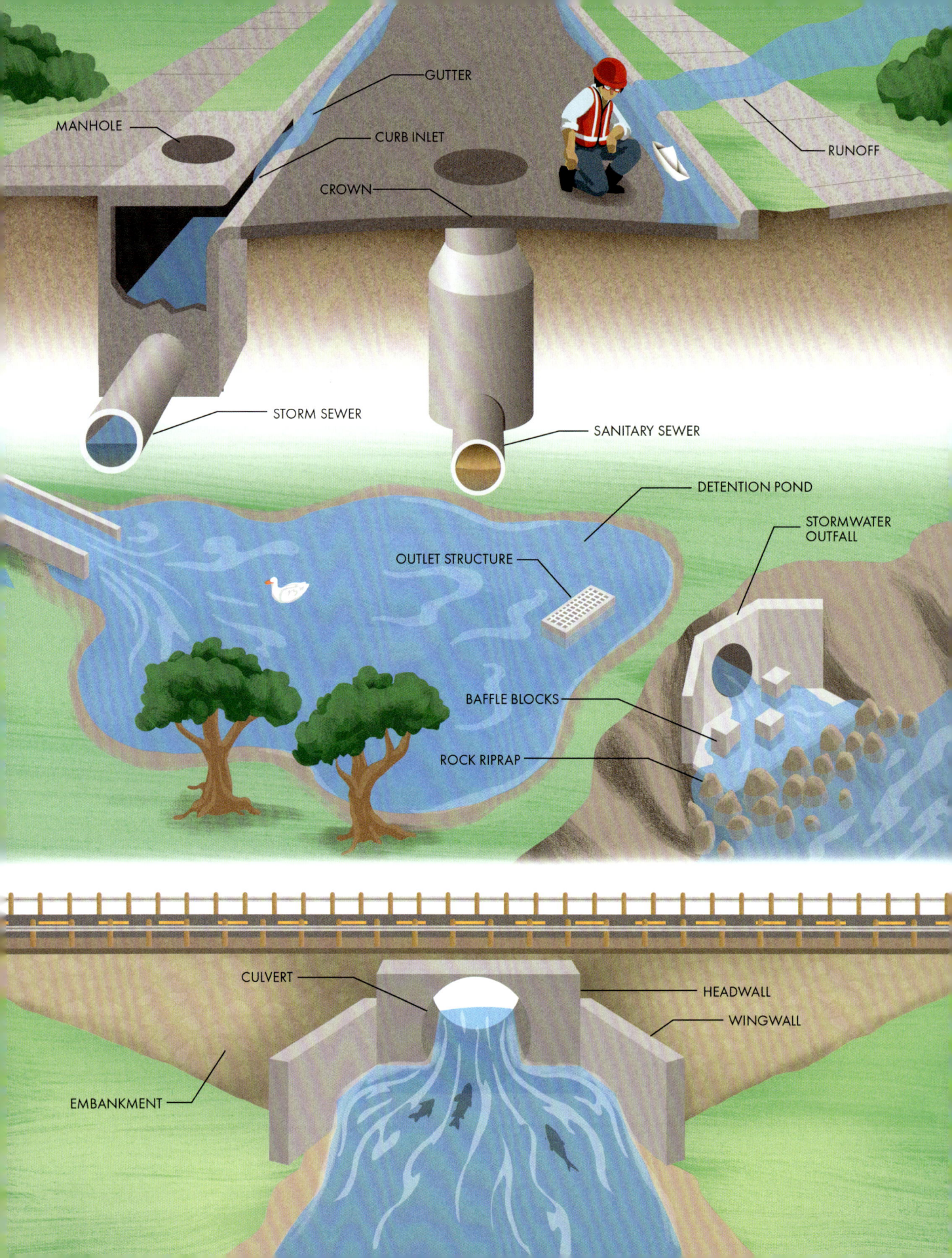

Stormwater Collection

One of the most significant environmental impacts of cities is how they affect the ways water moves above and below the ground during storms. All those streets, sidewalks, buildings, and parking lots cover the ground with impervious surfaces so that, instead of rainwater infiltrating, it runs off toward creeks and rivers, swelling them faster and higher and filling them with more pollution. Where a natural watershed acts like a sponge to absorb and slow rainwater as it falls, urban watersheds work more like funnels, gathering and concentrating runoff. Stormwater and flooding have been a problem ever since people started living in cities, and the first solution was simply to get the water out and away as quickly as possible. This solution is in the name we still use for how cities manage storms: *drainage*. When it rains and when it pours, we try to give that runoff somewhere to go.

Most cities are organized such that the streets serve as the first path of flow for rainfall. Individual lots are graded with a slope toward the road so that water flows away from buildings where it would otherwise cause problems. The standard city street has a CROWN in the center with GUTTERS on either side for water to flow. This keeps the road mainly dry and safe for vehicle travel while providing a channel to convey RUNOFF. Eventually, the road will reach a natural low point and start back uphill or will have collected so much runoff that it can't hold it all in the gutter.

It is possible to let runoff from a street outfall directly into a natural watercourse in some cases. However, in dense urban areas where space is limited, stormwater is often routed into underground drains.

In the past, it was common to simply put all the runoff from the streets directly into the sanitary sewage system. Unfortunately, wastewater treatment plants are usually not designed to process massive influxes of combined sewage and stormwater runoff at the whims of Mother Nature. In the worst cases, these plants have to release untreated sewage directly into watercourses when the inflow is too much to be stored or processed. That's why most cities now separate STORM SEWERS from the SANITARY SEWERS used to carry wastewater. Rain usually enters a storm sewer system through a CURB INLET or surface grate. Inlets are located at all low points along a road (called *sags*) and at regular spacing on sloped sections. Many inlets include a MANHOLE so that they can be accessed for cleaning and maintenance. A storm sewer pipe connects to each inlet to carry rainwater away. Each line is sized and sloped for gravity flow according to the expected volume of stormwater, similar to how sanitary sewer pipes are designed to carry a specific volume of wastewater.

A storm sewer system converges and concentrates just like a natural system of streams and rivers. Eventually, the sewers are routed to an OUTFALL in a natural watercourse or the ocean. Energy-dissipating BAFFLE BLOCKS or ROCK RIPRAP armoring are often installed at outfalls to protect the natural soils against erosion from the swiftly exiting runoff. Unlike a sanitary sewer system that terminates in a treatment plant, most stormwater runoff is discharged directly into the environment, so cities often include warnings to the public about disposing of waste in curb inlets.

Storm sewers help reduce local flooding by quickly removing water from streets, conveying it to streams and rivers. However, the influx of stormwater from urban areas exacerbates flooding within these natural water bodies. Many cities increase the capacity of natural watercourses by enlarging, straightening, and lining them with concrete. This design strategy is often known as *channelization*. Speeding up stormwater flow through channelization helps reduce the depth and extents of flooding, but it also has disadvantages. Dirty concrete channels hurt the visual character of a city. Channelization can also worsen flooding downstream and degrade the habitat of the original waterway. Most cities recognize that widening and lining natural channels is an incomplete solution to the increased runoff from urban development.

As a result, cities now require developers to take responsibility for their own impacts on stormwater volume and quality, which usually involves onsite storage before releasing drainage into a watercourse. *Retention ponds* maintain a permanent pool of water where DETENTION PONDS are ordinarily dry. Both act like mini sponges, absorbing all the rain that rushes off the buildings, streets, and parking lots. Their OUTLET STRUCTURES are designed to slowly release runoff back into waterways, shaving off the peak flow rate to the level before all the buildings and parking lots were built. Retention and detention ponds also help reduce pollution by slowing down the water, so suspended particles can settle out.

Along highways, it is not usually economical to manage stormwater below the ground. Instead, we often build highways above the natural ground on EMBANKMENTS with parallel ditches to carry stormwater. When a roadway crosses over a significant stream or river, we often construct a bridge. However, it isn't cost-effective to bridge over every minor channel and topographic depression in the landscape. When a roadway intersects with a minor watercourse, a CULVERT allows water to cross below from one side to the other. Engineers select the culvert pipe size to reduce the possibility of rainwater overtopping the roadway. HEADWALLS and WINGWALLS hold the embankment back while guiding stormwater into the culvert. A poorly designed culvert can let water through while obstructing the movement of animals, so engineers work with biologists and environmental scientists to ensure that culverts are appropriately designed for both the water they must carry and the creatures that live within it.

Municipal drainage infrastructure has come a long way, but it still primarily treats stormwater as a waste product, something to be gotten rid of. The reality is that rainwater is a resource, and natural watersheds provide many more services than simply conveying runoff downstream. They serve as habitat for wildlife, clean and filter runoff with natural vegetation, divert rain into the subsurface to recharge aquifers, and reduce flooding by slowing down the water at the source rather than letting it quickly wash away and concentrate. Many cities are moving toward ways to replicate and re-create natural watershed functions within developed areas. In the United States, practices that reduce the volume of runoff and the pollution it carries by managing it close to the source are collectively called *low impact development*. They include strategies like rain gardens, vegetated rooftops, permeable pavements, strips of vegetation used to filter surface runoff, and other ways to harmonize the built environment and its original hydrological and ecological functions. Low impact development can also include better floodplain management by using land for purposes that are less vulnerable to flooding, such as parks and trails.

8

CONSTRUCTION

Introduction

All infrastructure has one thing in common: it must be built. You can't buy a sewer system or electrical grid off the shelf at a store. Rather, these complex facilities are constructed in place by humans and machines. Construction can be both a nuisance and a joy, depending on your perspective (or your commute). It often seems noisy, disruptive, and slow. However, the giant-sized equipment and the urgency of effort evoke a sense of wonder and awe in the attentive observer. There is nothing quite like witnessing a structure take shape from raw materials and hard work, and it's often difficult to walk past any construction site without being distracted by the continual commotion.

Although construction can seem chaotic, there is a method to the madness. Each separate worker and piece of equipment has a specific task. The individual accomplishments may seem insignificant or even mundane, but they slowly accumulate into results that can be spectacular (as shown in the previous chapters). Observing construction sites can be a one-time activity of spotting machines and equipment or a regular routine to marvel at the steady progress. No matter how you choose to watch, you can always see something interesting on a construction site.

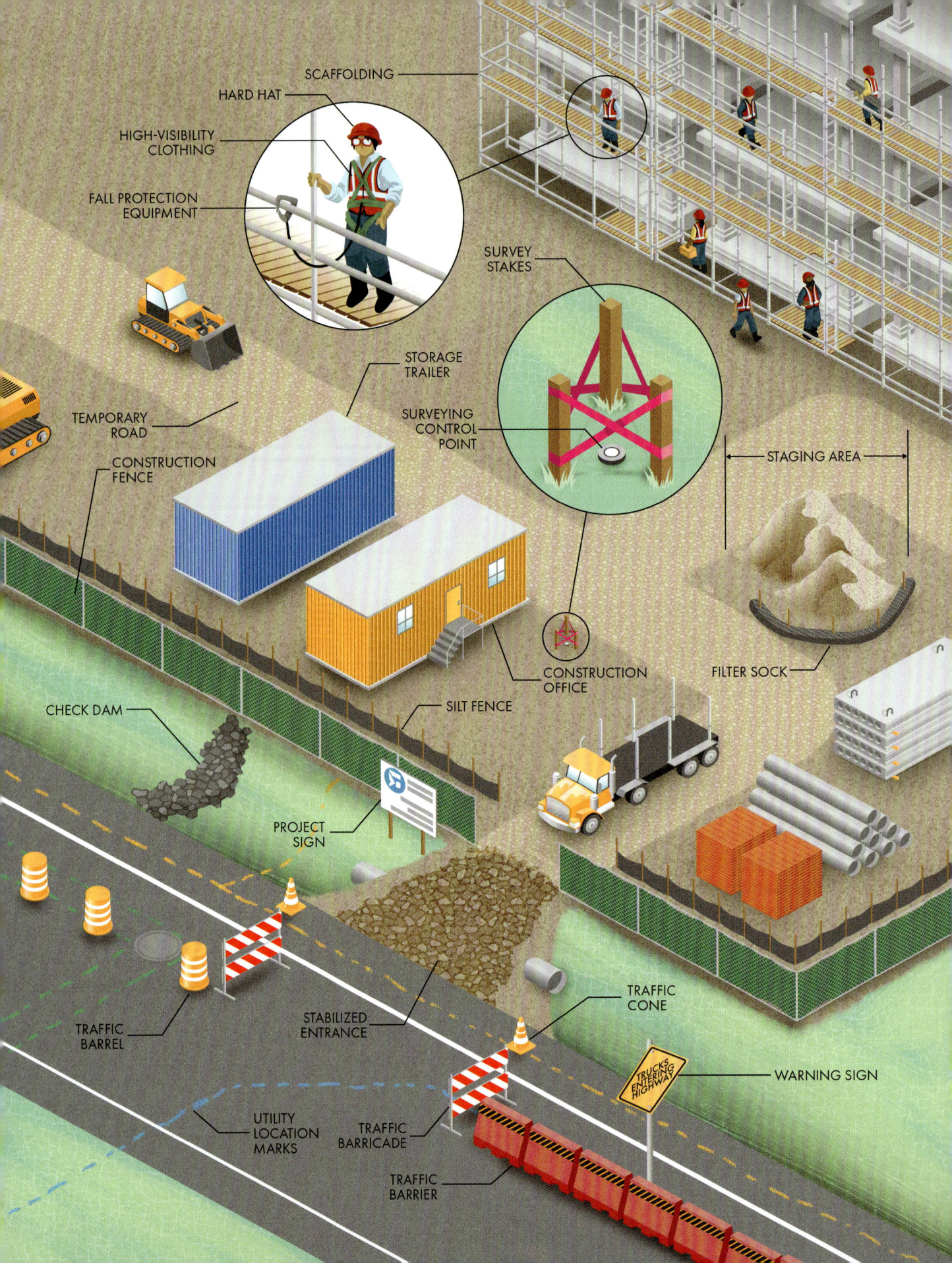

SCAFFOLDING

HARD HAT

HIGH-VISIBILITY
CLOTHING

FALL PROTECTION
EQUIPMENT

SURVEY
STAKES

STORAGE
TRAILER

SURVEYING
CONTROL
POINT

TEMPORARY
ROAD

STAGING AREA

CONSTRUCTION
FENCE

CONSTRUCTION
OFFICE

FILTER SOCK

CHECK DAM

SILT FENCE

PROJECT
SIGN

TRAFFIC
CONE

WARNING SIGN

TRUCKS
ENTERING
HIGHWAY

TRAFFIC
BARREL

STABILIZED
ENTRANCE

UTILITY
LOCATION
MARKS

TRAFFIC
BARRICADE

TRAFFIC
BARRIER

Typical Construction Site

Whether it's a road, bridge, dam, pipeline, or any other piece of infrastructure being built, a construction site may look like a disorganized frenzy of machines and activity at first glance. However, look closely, and you'll begin to understand the rhythm. Although every construction job is unique, the site where the work takes place is often remarkably similar from project to project.

Before construction can begin, a surveyor must lay out the location of the project on the ground. Surveyors install CONTROL POINTS away from the area of disturbance that can be used as a reference once construction begins. They are usually large nails driven into the underlying concrete or asphalt or iron rods driven into the soil. Surveyors often mark control points and other essential features of construction with wooden STAKES and plastic flagging tape. Linear construction projects like roadways and pipelines often use a coordinate system called *stationing*. In the United States, each station is equal to 100 feet. It's common to see locations on a site labeled with their distance along the centerline of the structure in stations plus the number of feet (for example, "STA 12+50" indicates a location 1,250 feet along the axis).

In addition to surveying, all subsurface utilities must be identified and marked to ensure that excavators don't inadvertently damage underground lines. Locators use colored spray paint to create UTILITY LOCATION MARKS on the ground. In many parts of the world, these colors are standardized.

For example, red is used for electric lines, orange for telecommunications, yellow for natural gas, green for sewers, and blue for drinking water lines. White paint is used to show the location of any excavations that will take place during construction, and pink is reserved for survey markings.

The first thing you might notice on a site is the PROJECT SIGN, which is used to identify the companies involved, notify the public of the name and purpose of the project, and post important information such as building permits.

Besides the structure itself, much of a construction site is usually dedicated to moving and storing materials. Heavy equipment and large trucks need space to move, load, and unload supplies. Having these large vehicles drive directly on the ground typically ends in a muddy mess, especially after rain. Therefore, contractors often construct TEMPORARY ROADS on the site to keep construction traffic flowing. In addition, most sites include a STAGING AREA for offloading and storing equipment and supplies to be used later in the project.

Although at a glance it might look like construction involves a lot of standing around, anyone who has worked in the trades can tell you that it's challenging work. Most of the people on a construction site are tradespeople who perform skilled labor such as masons, carpenters, welders, painters, and ironworkers. In addition, you may see a *superintendent* who oversees the project, an *inspector* who ensures that construction is performed according to the project plans and specifications, and safety personnel who act as spotters for potential accidents and resolve any hazards before they can result in injury.

Construction sites are particularly hazardous because of the large vehicles, the dangerous tools, and the need to work in precarious locations and heights. Many of the elements you can observe on a site are related to worker safety, including the *personal protective equipment* worn by each individual. Workers and other staff on site are usually required to wear a HARD HAT to protect against falling or protruding objects. Workers also wear HIGH-VISIBILITY CLOTHING featuring bright colors and reflective stripes to prevent accidents that result from not being seen. When working at heights, SCAFFOLDING is used to provide temporary platforms for crews to access challenging areas. Workers may wear FALL PROTECTION EQUIPMENT, including a harness and tether, to reduce the danger of falling while working at heights or near deep excavations.

In addition to keeping workers safe, construction projects must consider the safety of the public as well. Most sites will include a FENCE to keep wayward pedestrians out of hazardous areas, sometimes fitted with a screen to keep the wind from blowing dust and discourage theft by hiding expensive tools and equipment.

Public safety is essential for roadway projects, which often require the closure of traffic lanes or a detour around the site. Contractors install traffic CONES, BARRELS,

BARRIERS, and BARRICADES to redirect vehicles and keep them away from construction activities. WARNING SIGNS and barriers are always orange so that drivers can distinguish them easily from other signs to proceed with caution through the work zone.

Construction doesn't only involve hard labor and power tools. Like any other business, much of the work happens in an office, such as ordering materials, reviewing plans, holding meetings, and answering email. On large projects, contractors often have an entire workforce of office staff on site to support the construction and keep things running smoothly. You may see one or more trailers that serve as temporary CONSTRUCTION OFFICES for the contractor, site engineer, or owner to use when needed. Other trailers may be used for the STORAGE of tools and materials.

One nuisance created by construction occurs from disturbance of the ground. Rain can easily wash away unprotected soil. These suspended sediments are considered pollutants because they degrade the quality of natural water bodies and impact wildlife habitat. So, most construction projects are required to have facilities that control stormwater runoff and keep it from carrying soil off the site. SILT FENCES and FILTER SOCKS slow down runoff so the sediment can drop out of suspension. STABILIZED ENTRANCES use rocks to knock the mud off the vehicles' tires before they leave the site. Finally, CHECK DAMS are placed in channels to keep flows from concentrating, which reduces the potential for erosion.

Many types of infrastructure, including wharves, bridges, and dams, are founded below the water. Constructing foundations where humans and machines cannot efficiently work is a significant challenge. As a result, much of underwater construction involves getting rid of the water first so that work can proceed in the dry, a process called *dewatering*. Often, a structure called a *cofferdam* is required to hold water back from a construction site temporarily. Cofferdams usually consist of earthen or rockfill embankments, interlocking steel plates called *sheet piles*, steel frames with a plastic membrane, or water-filled rubber bladders. They aren't always completely watertight, so construction sites with cofferdams require pumps to keep the area dry. After construction is complete, the cofferdam is removed, flooding the site back to its original underwater state. For construction sites on rivers and canals, the dewatering process also requires flows to bypass the work. Depending on the volume, these bypasses may be handled by pumps, temporary channels, or tunnels, or by constructing a project in multiple phases while routing flows through the inactive part of the site.

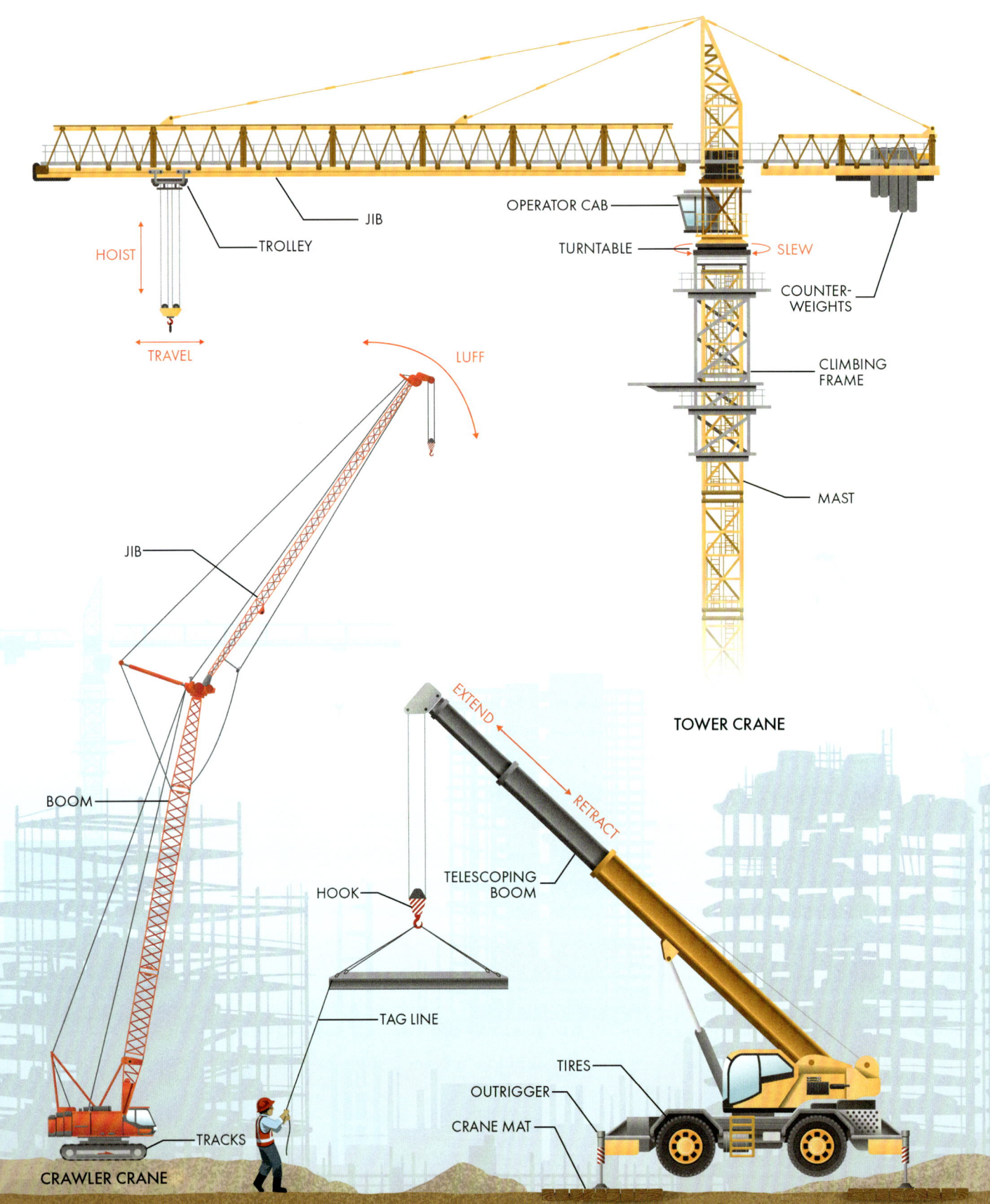

JIB

TROLLEY

HOIST

TRAVEL

LUFF

OPERATOR CAB

TURNTABLE — SLEW

COUNTER-WEIGHTS

CLIMBING FRAME

MAST

JIB

BOOM

EXTEND

RETRACT

TELESCOPING BOOM

HOOK

TAG LINE

TOWER CRANE

TIRES

OUTRIGGER

CRANE MAT

TRACKS

CRAWLER CRANE

ROUGH TERRAIN CRANE

Cranes

All construction can be boiled down to material handling: receiving deliveries, and storing, moving, and placing all the pieces and parts of a project. Of course, sweat and muscles can handle much of this work, but anyone working in the trades will tell you that there are plenty of jobs only a crane can accomplish. At many job sites, the question is not whether a crane will be used, but how many and what types. These backbones of the construction industry make it possible to lift and install materials and components much, much larger and heavier than would otherwise be possible with only human labor, making construction faster and more efficient than ever before.

Many types of cranes are used on construction sites, and each has its own benefits. They are usually divided into two types: *mobile* and *fixed cranes*. Mobile cranes have wheels or tracks, allowing them to move to different parts of a site. A CRAWLER CRANE is mounted on an undercarriage with a set of TRACKS. Crawler cranes are the largest and most capable type of mobile cranes found on a construction site. They are often equipped with steel BOOMS that can reach long distances and great heights. The largest booms are made from a lattice of steel bars to be both lightweight and extremely stiff. In addition, many manufacturers offer JIBS that can be attached to the end of a boom to extend its reach even farther. Crawler cranes aren't legal to drive on roadways, so they are usually trucked to a project to be assembled on site.

Like crawler cranes, ROUGH TERRAIN CRANES ride atop a mobile undercarriage, but they use rubber tires instead of tracks. Rough terrain cranes are best at accessing remote and challenging locations. They are usually smaller than crawler cranes, making them faster to set up and easier to fit into spaces that other cranes could not. Many rough terrain cranes feature TELESCOPING BOOMS with sections that extend outward to increase the crane's reach. They can be driven (slowly) while carrying a load, making it possible to move heavy objects across long distances on a construction site. However, their load rating increases significantly when operating in a fixed location on OUTRIGGERS. These devices stabilize the crane by lifting the undercarriage off the pliable tires. *All-terrain cranes* work and look similar to rough terrain cranes, but they are designed to drive on streets and highways, eliminating the need to be hauled to a job site on a truck. They are usually the smallest of the mobile cranes but also the most versatile.

Fixed cranes are installed in a single location where they remain for some or all of a project's duration. On a construction site, the most common type of fixed crane is the TOWER CRANE. It consists of a vertical MAST and a horizontal JIB extending out from the tower. The jib can rotate in any direction around the mast on a TURNTABLE. A TROLLEY that rides along the jib allows the operator in the overhead CAB to place the HOOK wherever needed.

Installation of a tower crane is a feat on its own, so they usually are used only for projects with extended durations, like tall buildings. They often have a base of reinforced concrete and require another crane for assembly and disassembly. Some tower cranes can raise themselves, allowing the mast to increase in height as a building is built up from the ground. A CLIMBING FRAME secures two sections of the mast as they are disconnected, lifting the upper part of the crane. Next, the crane raises and inserts a new mast section into the opening the climbing frame creates, where it is bolted into place. This process can be repeated as many times as needed to reach the desired height.

The primary goal of any crane is to relocate or reposition a load from one place to another, so cranes have many ways to move. Nearly all cranes have a drum onto which cabling is spooled. When a crane rotates the drum to lift a load using the cable, it is called HOISTING. In addition to hoisting, some booms can pivot, allowing the crane to change the boom angle with cables or hydraulic cylinders. When the boom of a crane pivots up with a load, it is called LUFFING. Some cranes can luff the boom and jib separately, providing an even greater range of movement. Swinging the boom or jib horizontally is often called SLEWING. Finally, cranes with telescoping booms can EXTEND or RETRACT, and tower crane trolleys can TRAVEL inward or outward.

Spotters on the ground communicate to the crane operator which movements are needed to attach, secure, lift, and place a load. If radios are not available, standardized hand signals let the operator know which type of movement the crew needs. Ground crews also use TAG LINES when necessary to control the load and keep it from spinning.

As vital as they are to construction sites, cranes can also be dangerous. Lots of engineering is used to keep them from tipping over. Wooden pads called CRANE MATS are often used to distribute the extreme pressure of cranes and keep them from sinking into the ground. COUNTERWEIGHTS made from steel or concrete are used to balance the load on the hook, reducing the crane's tendency to tip over (called its *moment*). Finally, on windy days, mobile cranes are generally taken down, and tower crane brakes are released so that the jibs can *weathervane*, allowing them to rotate freely with the wind rather than fight against it.

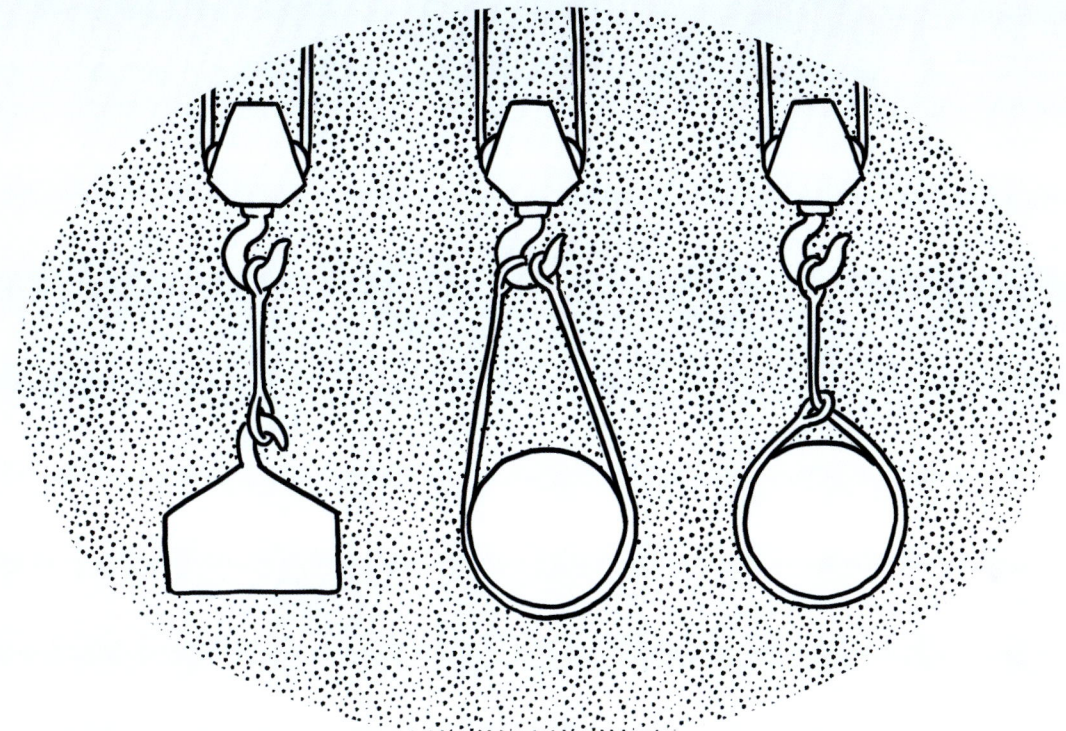

Most cranes use a hook to connect to the load, yet few things that need to be lifted have an attachment that fits nicely over a gigantic steel hook. *Rigging* is the term used to describe all the steps we go through to attach a load to a crane so it can be suspended and moved. A rigger's most commonly used tool is a *sling*: a simple length of cable, chain, rope, or webbing with eyes or hooks on either end. Slings are generally used in one of three basic *hitches*. In a vertical hitch, one eye of the sling is connected to the hook, and the other is connected to an attachment point on the load. In a basket hitch, the sling passes under the load with both eyes on the hook. Finally, in a choker hitch, one eye of the sling is wrapped around the load, passed through the other eye, and attached to the hook. Each hitch used in rigging comes with a different load rating and specific advantages over the alternatives. Next time you watch a crane lifting a load, see if you can spot which of the three hitches is being used with the slings.

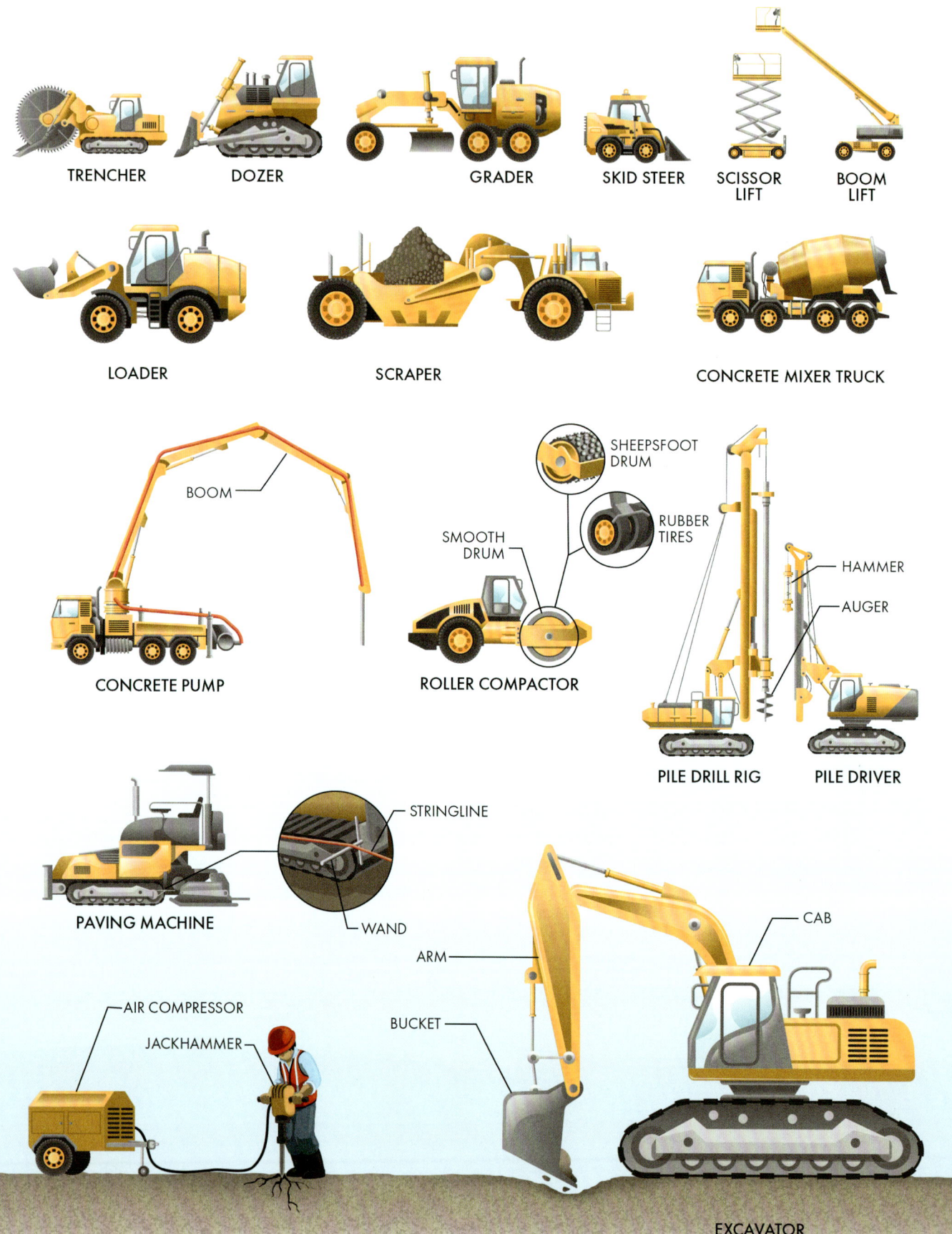

TRENCHER

DOZER

GRADER

SKID STEER

SCISSOR LIFT

BOOM LIFT

LOADER

SCRAPER

CONCRETE MIXER TRUCK

BOOM

CONCRETE PUMP

SHEEPSFOOT DRUM

SMOOTH DRUM

RUBBER TIRES

ROLLER COMPACTOR

HAMMER

AUGER

PILE DRILL RIG

PILE DRIVER

STRINGLINE

WAND

PAVING MACHINE

AIR COMPRESSOR

JACKHAMMER

ARM

BUCKET

CAB

EXCAVATOR

Construction Machines

Nothing amplifies human effort more than heavy equipment. In addition to the cranes described in the previous section, a myriad of machines is used on construction sites to increase the speed and efficiency of work. Although construction machines may seem like just a cacophony of hammering and backup alarms, modern buildings and infrastructure could not exist without their capabilities to move, compact, dump, drill, deliver, break down, and build. Of course, describing every piece of equipment one might observe on a construction site would be impossible, but you're bound to the ones described here if you keep an eye out.

A large number of construction machines are meant for performing earthwork: moving and placing soil and rock. EXCAVATORS can be found on many construction sites because of their versatility. They typically feature a BUCKET, ARM, and rotating CAB, although many other attachments and configurations exist. Excavators use their hydraulic cylinders to perform various functions, including digging holes and trenches, removing debris, and even lifting and placing loads like a crane. They come in many different sizes, from mini excavators that can fit in the back of a pickup truck to gigantic machines too large to be transported on a highway in one piece. TRENCHERS are another type of machine meant specifically for excavation. Trenchers use a toothed wheel or chain to cut long holes into the earth for installing pipelines, drains, electrical lines, and other linear utilities.

DOZERS feature large blades used to shove materials around. They can clear brush, trees, and boulders from a site, push soil across short distances, and spread fill over a large area. Like dozers, GRADERS also feature a long blade. However, graders offer more precision, allowing operators to level (or *grade*) the earth with high accuracy. Instead of a blade, LOADERS are equipped with large buckets for excavating and transporting larger quantities of soil. These machines also come in many sizes, from the tiny SKID-STEER loaders used on small sites to the massive wheel loaders used in mines. When significant volumes of soil must be moved across a site, SCRAPERS are the machine for the job. Scrapers cut into the earth like a carpenter's plane, filling a hopper on the vehicle. Then the soil can be transported and placed directly from the scraper, eliminating the need to transfer it to another vehicle like a dump truck.

Many construction machines are dedicated exclusively to performing roadwork. PAVING MACHINES are used to lay asphalt on roads, bridges, and parking lots and create concrete curbs, gutters, barriers, and highways. Dump trucks or loaders feed asphalt or concrete into the machine, which uses a series of mechanical devices to create a smooth and even layer as it travels. Pavers that make concrete structures use a process called *slipforming* to create continuous shapes like curbs and highway barriers. Because the subgrade isn't always perfectly level, many paving and slipform machines use a WAND that rides along a STRINGLINE set by surveyors according to the engineered alignment of the roadway. The wand controls the steering and position of the paving machine to ensure that the road and its other features are smooth and consistent.

Concrete structures cure and harden on their own, but earthen materials and asphalt pavement must be compacted into place. On large projects, this task is typically accomplished using a ROLLER COMPACTOR, a heavy vehicle with one or two SMOOTH DRUMS that compress the soil or asphalt as they roll over it. Some roller compactors are equipped with RUBBER TIRES that help knead the subgrade or asphalt, speeding the compaction processes. Similarly, compactors used on sticky clay soils sometimes use a textured roller called a SHEEPSFOOT DRUM. Finally, many roller compactors can vibrate, increasing their ability to smooth and flatten the surface.

Piles are another common part of many construction projects that consist of vertical structural elements drilled or driven into the ground to create retaining walls and foundations. PILE DRIVERS use large HAMMERS or vibratory mechanisms to push steel or concrete piles into the subsurface. Where piles are installed in holes, a DRILL RIG is used to excavate shafts using an AUGER or other rotating tool.

Many construction projects involve casting wet concrete into forms on site. You may have seen a CONCRETE MIXER TRUCK transporting concrete from a plant to a project site. The large drum has a spiral-shaped blade inside. When the drum rotates in one direction, it mixes concrete ingredients to keep them from separating during transit. When the drum rotates in the other direction, concrete is pushed toward the back of the drum to be discharged. Some projects can use concrete directly from a truck, but often concrete is needed in locations that are difficult to access. Therefore, mixer trucks often discharge into CONCRETE PUMPS that can transport the wet mix through pipes. Some concrete pumps are also equipped with an articulating overhead BOOM, allowing concrete to be placed wherever needed with high accuracy.

Another class of construction machines called *aerial lifts* has the simple goal of safely positioning workers in precarious locations. A SCISSOR LIFT consists of a mobile base driven by an operator and a platform that rises vertically atop crisscrossing braces. They move only vertically, so scissor lifts can't be used to maneuver workers around obstacles. A BOOM LIFT uses a hydraulic arm to support the platform, offering more freedom to access difficult areas on a construction site.

Of course, in addition to the many vehicles used on construction sites, workers also use a litany of powered hand tools. Some of the most important construction tools are *pneumatic* (in other words, powered by air), necessitating an AIR COMPRESSOR on site. It is common to spot trailer-mounted air compressors on construction projects used to power JACKHAMMERS, drills, grinders, nail guns, and many other tools workers use.

The capabilities of construction equipment are rapidly increasing with newer and more advanced technologies. One innovation that has completely changed the world of earthwork is the Global Positioning System (GPS). There isn't much need for map navigation on a construction site, but GPS technology offers many applications beyond the way we normally use it in our cars. GPS devices know where a machine is located on the site and the position of its tool in relation to the *final grade* (the desired elevation of the ground). Traditional projects require surveyors to meticulously stake out the locations and extents of earthwork, sometimes multiple times throughout a job. GPS-enabled equipment uses a digital model of the project and an onboard interface to show the operator exactly where to guide the machine. In some cases, the GPS device can even control the blade or bucket automatically. Many systems have multiple circular antennas mounted to the machine, so it's easy to notice if the equipment is taking advantage of GPS.

Acknowledgments

I owe a great deal of thanks to those who made this book possible:

My wife, Crystal, who is endlessly supportive, occasionally funny, and the love of my life.

My son, Cliff, who unwittingly pushed me to choose a career *about* engineering over one in it.

My brother, Graham, who showed me how to take risks, is the sounding board for every idea I have, and is my most helpful critic.

My cousin, Samuel, who was the first participant of "What's That Infrastructure?"— a road trip game I invented so that I could talk about civil engineering more.

My best friend and collaborator, Wesley Crump, who first suggested that I write a book and then became an invaluable team member when I decided to actually do it.

My parents, Joe and Carol, whose support, encouragement, and exemplification of all of life's important skills eventually led me to where they knew I was headed all along.

My editor, Jill Franklin, and all the staff at No Starch Press who immediately understood my vision, patiently coached me through the book writing process, and worked terribly hard to create something special.

My illustration team at MUTI, led by Brad Hodgskiss, who converted my conglomerations of stock images and chicken scratches into imaginative works of art.

The technical reviewers listed at the beginning of the book who provided their wisdom and experience to improve each chapter and catch my errors.

All my professors at Texas State University and Texas A&M University and all my former colleagues at Freese and Nichols who shared with me their enthusiasm and expertise in engineering, construction, environmental science, and much more.

Finally, all the fans of *Practical Engineering* on YouTube and elsewhere for your comments, email, and viewership. I could never have written this book without the encouragement and feedback I've received from you over the past six years.

Glossary

NUMBERS

100-year flood The flood magnitude that has a 1 percent chance of being equaled or exceeded in a given year.

A

abutment The structure or geologic formation that forms the end of a *bridge* or *dam*.

access hatch A door into a restricted area.

accessibility The design of structures and environments to be usable by people with disabilities.

accretion The growth process of a *bank* or shore by gradual accumulation of sediments.

activated sludge Aerated *microorganisms* used to remove nutrients from wastewater.

active device A device that requires an outside source of power to operate.

active warning device A device that provides advance notice of the approach of a train.

actuated signal control A *traffic signal* control scheme that uses vehicle detectors in setting the time for each *phase*.

adaptive signal control technologies (ASCT) A *traffic signal* control scheme that uses sensors to set the timing of individual signals based on conditions within the larger traffic network.

administrative building In the context of power generating stations, a building used to house offices for administrative employees, including engineers.

aeration basin A storage structure at a *wastewater treatment plant* used to introduce dissolved oxygen to the sewage.

aerial lift A machine used to position workers in high or difficult locations.

aerobic In the presence of oxygen.

aggregate A material consisting of coarse- to medium-sized rock particles, including *sand* and *gravel*.

air compressor A machine that increases the *pressure* of ambient air, often for the purpose of powering construction tools or equipment.

air insulated switchgear Switches, fuses, breakers, and other equipment used in electric power stations and *substations* that rely on open air for insulation.

air lock A restriction or blockage of flow in a pipe caused by a trapped bubble of vapor.

air release valve A *valve* that releases air from a liquid *pipeline*.

alignment The horizontal *layout* of a roadway when looking from above.

all-terrain crane A *mobile crane* that can drive on a highway and access on-road and off-road construction sites.

alternating current (AC) Electric *current* that reverses directions periodically.

altitude The vertical distance between a reference surface and an object.

amplifier A device that increases a signal's strength.

AM radio Transmission of information by radio waves where the strength of the signal is varied in proportion to that of the message.

anaerobic In the absence of oxygen.

anchor A device for affixing a structure to the earth.

anchorage A formation of rock or *concrete* into which *anchors* are installed.

angle of repose The steepest angle at which a pile of granular material will rest without collapse.

annular space The space between two cylindrical structures placed inside each other.

antenna A device that serves as the interface between radio waves and electric signals.

antenna array A group of connected *antennas* that work together to send or receive signals directionally.

antenna level A vertical location along a *cell tower* at which a single service carrier's *antennas* are mounted.

approach The area of transition between a *bridge* and roadway.

aqueduct A structure designed to convey water over long distances (sometimes referring specifically to a *bridge* that conveys water over a valley).

aquifer An underground *reservoir* of water.

aquitard A geologic *formation* that slows or stops the flow of groundwater.

arc A breakdown of air between two *electrodes* that allows electric *current* to flow, often visible as a bright discharge.

arch A curved structural element used to support a load across a gap.

arch bridge A *bridge* that uses curved structural shapes to transfer loads to its *abutments*.

arch dam A curved *dam* that transfers reservoir *pressure* to its *abutments*.

armoring See *revetment*.

armor unit A *precast concrete* structure used to protect a shore or *bank* against erosion.

arrester A protective device that routes energy to ground during an electrical surge.

arterial roadway A high-capacity urban road that connects *highways* to *collector roads*.

artificial reef An artificial structure installed to promote marine life.

asphalt A durable pavement material made from *aggregate* and *bitumen*.

at grade At ground level.

automated guided vehicle (AGV) A driverless robot used to transport cargo around a *yard* or industrial facility.

auxiliary spillway A secondary *spillway* designed to be used infrequently and only under extreme flooding conditions.

B

backfill Soil or rock replaced into an excavated area.

backhaul The part of a mobile telephone network that connects individual *base stations* to the network backbone.

backup generator A device that provides electric power if grid power is lost, usually powered by a gasoline or diesel engine.

backwash The process of reversing the flow of a fluid through a *filter* to clean the media.

baffle block A structure used to dissipate kinetic energy in a stream of flowing water.

baffled chute A *chute* or *spillway* with an array of *baffle blocks* to limit the flow velocity as it operates.

baghouse An air pollution control device that removes particulates using fabric bags as filters.

ballast *Aggregate* material used to transfer the load of *railroad tracks* to the *subgrade*.

bank The sloped land along a river or lake.

bank intake A water *intake* structure installed on the *bank* of a river.

barrel A warning device used on roadways to separate construction areas from *travel lanes*.

barricade A warning device used to bar entry to vehicles.

barrier A warning device used to separate traffic streams and protect areas from errant vehicles.

barrier gate A device that bars vehicle entry when closed.

bar screen A coarse mesh of metal bars used to catch trash and debris in a stream of water.

base A layer of compacted material below the wearing surface of a roadway that provides structural support.

base station A site where *antennas* and communications equipment are placed to create one or more *cells* in a cellular network.

basket screen A *bar screen* formed into the shape of a box or basket.

beach nourishment The process of replacing sediment on a beach to fight erosion and increase beach size.

beam (naval architecture) The width of a ship or boat at its widest point.

beam (structure) A linear structural element that spans a distance.

beam bridge A *bridge* that uses horizontal structural elements to span a gap between two *piers* or *abutments*.

bearing The resting surface between a bridge's *superstructure* and *substructure*.

bearing block A structural feature that distributes the force of an *anchor* into a wall or facing.

bedding A layer of *gravel* installed below a *revetment* to prevent erosion beneath the armoring layer.

bell A shaped area at the end of a pipe into which the *spigot* of another pipe fits to connect the two.

bending force A force applied perpendicularly to the long axis of a structural element.

bent A rigid frame used as an intermediate support for a *bridge*.

bentonite clay A very fine soil often used in drilling fluid and as a groundwater barrier in underground construction.

berm See *verge*.

Betz limit The theoretical maximum power that can be extracted from wind using a *turbine*, equal to approximately 59 percent of the wind's total kinetic energy.

bicycle lane A lane on a roadway dedicated for bicycle traffic.

biofouling The unwanted accumulation of aquatic organisms on a structure or vehicle.

biogas The flammable byproduct of *anaerobic* decomposition consisting of methane and other gases.

biosolid The solid byproduct of *anaerobic* digestion of *sludge* from the wastewater treatment process.

bird decoy An imitation of a predatory bird used to discourage real birds from perching nearby.

bird spikes Devices that obstruct potential perches to discourage birds from perching in unwanted locations.

bitumen A viscous mixture of hydrocarbons used as the binder in *asphalt*.

black hole effect A sharp transition in lighting at the entrance of a *tunnel*.

blackout A power outage resulting in total loss of electrical power to end users.

bladder A flexible sac inflated by water or air.

blade The individual elements that interact with wind to drive a *turbine*.

blasting hole A drilled hole in rock into which explosives are placed.

block A length of *railroad track* that may be occupied by only one train at a time.

boiler A vessel used together with a *furnace* to create steam from liquid water.

bollard A post on a *wharf* or dock to which ships are moored.

boom (machinery) The lifting arm of a *crane*, *excavator*, or other construction machine.

boom (water) A string of flotation devices used to warn or exclude people and boats from a dangerous area.

boom lift An *aerial lift* that uses a *boom* arm to position workers in high or difficult locations.

booster pump A machine used to increase the *pressure* of fluid in a pipe.

borehole A circular excavation into the earth created by drilling.

boring The act of drilling a hole into the earth.

box girder A structural *beam* that forms an enclosed tube.

branch sewer A sewer line that collects wastewater from *lateral sewers* and flows into a *main sewer*.

breakaway The property of a signpost or other obstacle designed to yield on impact to reduce the chance of injury.

breakwater A barrier installed to dissipate wave energy offshore to protect a *harbor*.

breather switch A diagonal *expansion joint* used on railroad *rails*.

bridge A structure that carries a road, path, or railway over a river or obstacle.

brownout A drop in *voltage* of an electrical power supply, usually due to a disruption or overloaded condition on the *electrical grid*.

bucket The part of a construction machine used to scoop and dump materials.

buffer Space provided between a bicycle and vehicle lane to provide additional comfort and safety to cyclists.

bulk carrier A ship that carries bulk goods such as *coal* or grain.

bundle A group of parallel *conductors* at the same electrical potential used to reduce *corona discharge* and increase capacity compared to a single large conductor.

buoy A floating device used to provide navigational information or warnings to ships.

bus A conductive element used to make electrical connections among various equipment in a *substation*.

bushing A hollow *insulator* that allows an electrical *conductor* to pass through a metal *casing*.

buttress A projecting support member along a wall or *dam*.

buttress dam A *dam* supported along the downstream face by a series of *buttresses*.

C

cab The part of a construction machine in which the operator sits.

cable-stayed bridge A *bridge* that uses diagonal cables from one or more vertical *towers* to support the weight of the *deck*.

cable termination A transition in an electrical *conductor* between a bare cable used in overhead lines to an insulated cable used in underground applications.

cable TV (CATV) A *telecommunications* network that uses *coaxial* or *fiber-optic cables* to deliver television and internet service to individual customers.

cable vault An enclosure that provides access to underground cables.

call button The button used at a pedestrian crossing to notify the *signal* that a pedestrian is waiting.

canal An artificial *channel* used for navigation or water conveyance.

cant See *superelevation*.

cantilever An overhanging structural element supported on only one side.

cantilever bridge A *bridge* that uses structures or structural elements that project horizontally to span a gap while only being supported on one end.

cap A structural member that transfers the loads from a bridge's *superstructure* to one or more *piers*.

capacitance The tendency of a *conductor* to store electric charge when subjected to a difference in electric potential.

casing The outer support pipe used in a *well* to keep the *borehole* from collapsing.

catenary The curved shape a wire or rope takes between two supports, or the system of overhead electric lines used in some electric railways.

cat's eye See *glass beads*.

CATV power supply A device that provides power to remote *amplifiers* on the *CATV* network.

cell The geographical area covered by a single frequency emitted by a base station.

cell on wheels (COW) A mobile cellular *base station* used to increase network capacity at large events or during emergencies.

cell site See *base station*.

cellular communication A *telecommunications* network that enables wireless *telephone* and internet using *base stations*.

cement grout A material consisting of cement and water used to seal areas or fill voids.

centripetal force The force required to move a body in a circular motion.

chamber See *lock*.

channel An excavated or natural linear depression in the landscape that conveys water.

channelization The process of straightening, widening, and lining a natural stream or river to increase its hydraulic capacity.

check dam A construction of stones in a *channel* used to slow down *runoff* and reduce its sediment load.

check valve A *valve* that allows flow in only one direction.

chicane An artificial *curve* added to a roadway as a *traffic calming* measure.

chute A sloped *channel* that carries water, usually made from *concrete*.

circuit In the context of electric *transmission lines*, an arrangement of three *conductors* corresponding to the three *phases* of a typical electrical power network.

circuit breaker A protective device that interrupts the flow of electrical *current*.

clarifier A circular pond used to settle solids from suspension that usually includes a *sludge* collection system.

Clarke Belt The line above Earth's *equator* where *geostationary* satellites orbit.

clearance The time period while a traffic *signal* is yellow and vehicles are cleared from the *intersection*.

clearwell A ground-level storage tank for finished water at a *water treatment plant*.

clear zone An unobstructed area along a *highway* that provides errant vehicles a safe space to stop.

climbing frame A device used on *tower cranes* to connect the upper and lower parts of the *mast* so that a new segment can be added.

clip A device that affixes a *rail* to a *tie*.

closure A gap to allow a roadway, railway, or path through a *floodwall* or *levee* that must be shut in advance of a flood.

closure rail The rail between the *point* and the *frog* in a track *switch*.

cloverleaf interchange A grade-separated highway *interchange* where all turns use right-hand ramps.

coagulant A chemical that neutralizes charges in suspended particles so that they can clump together.

coal A mined material that consists of carbonized vegetable matter.

coating A protective layer of paint or other material.

coaxial cable An electrical cable used to transmit high-frequency signals consisting of an inner *conductor* surrounded by an outer conducting shield.

cofferdam A structure used to temporarily impound water during construction.

collector pipe A pipe used in a subsurface *drain* to gather and discharge *seepage*.

collector roads The lowest capacity urban roads that connect individual houses and businesses to *arterial roadways*.

colocation The sharing of a single *radio antenna tower* or mounting structure by multiple service providers.

column A vertical structural element that supports loads from above.

column pipe The vertical pipe that carries water from a *well* or *sump* to the surface.

communications cabinet An enclosure for communications equipment to protect against weather and vandalism.

communication space The lowest space on a *joint pole* where *telecommunications* lines are attached.

complete streets Urban roads designed to enable safe use by all modes of transportation and user ability.

composite tank A *water tower* that uses a concrete *pedestal* and elevated steel tank.

concrete A mixture of cement, *aggregates*, water, and other admixtures that forms a solid, durable mass.

concrete dam A *dam* built primarily from *concrete*.

concrete mixer truck A truck equipped with a large barrel for mixing and delivering *concrete*.

concrete pump A device that pumps *concrete* used in locations where a mixer truck cannot access.

conductor An object or type of material that allows the flow of electric *current*.

conduit A pipe or other tubular construction for flowing water.

conduit riser A vertical tube used to protect electrical *conductors* that run along a *utility pole*.

cone A warning device used to mark temporary traffic control zones during construction.

connecting rod The rod that connects track *points* to the *switchstand*.

conservator A tank that provides space for thermal expansion of oil inside an electrical transformer.

constellation A group of *satellites* arranged in orbits to increase their coverage area.

construction office A building or trailer used to conduct business and hold meetings on a *construction site*.

construction site An area on which the activities associated with construction of a building or structure take place.

contact wire The wire in an overhead electric *catenary* system that provides power to the train's *pantograph shoe*.

container A large reusable box used in *shipping* of freight.

containment building An airtight building around a nuclear *reactor* meant to contain the escape of radioactive gas in the event of an emergency and protect the facility against attack.

control building At a *substation*, a building that houses *relays*, controls, batteries, communications gear, and other low-voltage equipment.

control joint A *joint* that artificially weakens a *concrete* slab to control where *cracks* form.

controlled-access roadway A high-capacity road where ingress and egress can only occur in select locations to minimize interruptions in traffic flow.

control point A marked point on a *construction site* used as a reference for horizontal and vertical dimensions.

conveyor belt A device that moves materials by means of a driven flexible belt.

cooling tower A device used to remove heat from a stream of water.

core (embankments) The central section of an *embankment* often constructed using a low-permeability material such as clay.

core (transformers) A conductive element that serves as the main path of the magnetic circuit within an electric *transformer*.

corner casting A feature at each corner of a shipping *container* that includes holes for securing and holding.

corona discharge The ionization of air surrounding a *conductor* carrying a high *voltage*.

corona ring A conductive ring used to distribute the electric field gradient on high-voltage *conductors* to reduce *corona discharge*.

countdown timer A pedestrian display showing how much time is available to cross before a traffic *signal* changes.

counterweight A weight used to balance another force or weight in a structural system.

coupler A device that connects cars together in a train.

course An individual layer of material used to construct a roadway.

cover A steel plate above a *manhole* used to keep people and debris out while allowing vehicles to pass over the top.

crack A structural separation of a material where both sides are still adjacent to each other.

crane A machine used to lift, move, and position heavy objects.

crane mat A construction of timbers used to distribute the weight of a vehicle to the ground.

crash cushion A device that absorbs vehicle impact forces to reduce the severity of a crash.

crashworthy The ability of a traffic control device to withstand a crash without posing undue hazard to vehicle occupants.

crawler crane A crane whose undercarriage rides on a set of *tracks*.

crest The top of a *dam* or *embankment*.

crest curve A vertical curve connecting two inclined sections of roadway at a high point.

crest gate A *gate* for controlling water that hinges at its bottom so that the top of the gate can change in elevation.

crib intake A large offshore structure that collects water from a lake, transferring it to shore via a *tunnel*.

cross arm An element fastened at a right angle to a *utility pole* to support power lines, *utility lines*, or other equipment.

crossbuck A *traffic sign* that indicates a level railway crossing.

crossing bell A device at a railway *grade crossing* that sounds to warn of an approaching train.

cross section The shape of a structure if it were cut along a plane.

crosswalk A designated and marked area for pedestrians to cross a roadway.

crown A *cross-sectional* roadway shape that is highest in the center with downward slopes to each side.

crusher A device used to reduce the size of bulk materials such as *coal*.

cul-de-sac A dead-end road that often includes a circular area so vehicles can turn around at the end.

culvert A *conduit* that carries *drainage* below a road or *sidewalk*.

curb A raised edge along a roadway that often forms a *gutter* to convey *drainage*.

curb cut A ramp connecting a *sidewalk* to the road surface through the *curb*.

curb inlet An opening along a roadway *curb* for *drainage* to enter a stormwater collection system.

curb strip See *verge*.

current The movement of charged particles through an electrical *conductor*.

current transformer A type of *instrument transformer* that scales large values of *current* to small values that can be measured using instruments and *relays*.

curve A bend in a roadway to bring a change in direction.

cut An area of excavation in an *earthwork* project.

cut-and-cover tunnel A *tunnel* installed inside an excavated *trench* from the surface.

cutoff wall A subsurface feature installed below a dam to reduce the volume and *pressure* of *seepage* in the foundation.

cutterhead The device at the front of a *tunnel boring machine* that rotates to abrade and remove material.

cuttings The material excavated from a *borehole*.

cylinder A steel tank used to store gas.

D

dam A structure built to impound water and create a *reservoir*.

damper (air handling) A device that regulates the flow of air in a *duct*.

damper (vibrations) A device used to reduce mechanical vibrations.

dead-end A *pipeline* that has a connection at only one end.

dead-end tower See *tension tower*.

deck The driving surface of a bridge's *superstructure*.

deck arch bridge An *arch bridge* where the *deck* is supported on the top of the arch.

deck truss On a *truss bridge*, a truss that runs below the road *deck*.

dendritic Of branching or converging form.

design speed A selected speed used to design geometric features of a roadway.

design vessel The maximum ship size used to select dimensions and features of a maritime facility.

detention pond An artificial pond that is normally dry and is created to temporarily store rainwater *runoff* to reduce flooding.

dewatering The act of impounding, bypassing, or removing water so construction or maintenance can be carried out in the dry.

diamond An assembly that allows two railways to cross each other.

diamond interchange A grade-separated junction between a *highway* and *minor road*.

dielectric oil Nonconductive liquid used as an *insulator* and coolant in electrical power equipment.

differential A gear system that allows driven wheels to turn at different speeds.

diffuser A perforated device used to introduce gas bubbles to a pool of liquid.

digestate See *biosolids*.

digester A vessel used in the wastewater treatment process to facilitate *anaerobic* decomposition of *sludge*.

digital subscriber line (DSL) A communications technology that allows transmission of digital data over a *telephone* line.

dike See *levee*.

dipole antenna An *antenna* that consists of two identical conductive elements, each connected to one side of the feedline.

direct current (DC) Electric *current* that is one-directional.

directional antenna An *antenna* that sends or receives signal at greater strength in a particular direction.

directional boring A method of installing underground utilities along a prescribed path without the need for *trenching*.

directional drill rig A machine that creates a horizontal *borehole* for installation of underground utilities.

direct potable reuse The process of treating wastewater to drinking water–quality standards and introducing it directly into a *potable* water supply.

discharge line The pipe on the downstream side of a *pump*.

disconnect switch A device used to de-energize equipment or *conductors* for repairs or maintenance but not usually intended to interrupt lines carrying significant *currents*.

disinfection The process of deactivating bacteria and other *microorganisms* that could cause disease.

distribution (electric power) The final stage in electric power delivery that carries electricity from *transmission* system to individual end users.

distribution transformer A *transformer* that reduces voltage used in *distribution* lines to the final level required by the end user.

ditch A small *channel* used to carry *drainage*.

ditch lights Lights below the main headlight on a train engine used to increase visibility at *grade crossings*.

Doppler shift A change in frequency that occurs when an observer is moving relative to the wave source.

downconversion The conversion of a high-frequency signal to a lower frequency to simplify transmission and processing.

dozer A machine equipped with a large blade for pushing material.

draft The depth to the lowest part of a ship's hull below the water line.

drain A device that collects and diverts water.

drainage See *runoff*.

dredge A machine used to remove earth from the bottom of a river, lake, or ocean.

drilled concrete shaft Structural *piles* created by placing *concrete* and reinforcing steel within a drilled *shaft* into the earth.

drill rig A machine used to drill into the earth.

drill string The assembly of pipes or shafts that transmit *torque* to the drill bit.

drop (electric railways) The support wire that connects the *contact wire* to the overhead *catenary wire*.

drop (telecommunications) A connection between the *telecommunications* network and an end user.

duct A pipe or *conduit* installed underground in which *telecommunications* lines are run.

E

earthfill Material used in construction that consists of any combination of sand, silt, or clay.

earth wire A wire used to connect a utility pole or piece of equipment to the ground; used as a safety measure to protect against shocks.

earthwork The act of excavating and filling areas with earth to reshape the landscape as part of a construction project.

effluent The liquid product of a treatment process.

elastomeric bearing pad A pliable rubber material that connects a bridge superstructure to the substructure while allowing some flexibility between the two.

electrical grid A network of interconnected power producers and users, usually spanning a large area.

electric motor A device that converts electric power into rotational motion.

electrified railway A railway on which the trains are powered externally by electricity.

electrode An electrical *conductor* that connects to a nonmetallic medium such as soil or air.

electromagnetism The interaction between electrically charged particles and magnetic fields.

electrostatic precipitator A device that removes small particles from the air using an electric charge.

elevated storage The act of storing water above ground level to maintain *pressure* within the distribution system and provide an emergency water supply.

embankment A linear area of earth *fill* used to carry a roadway or impound water.

embankment dam A *dam* constructed using *earthfill* or *rockfill*.

emergency exit A route from a building or *tunnel* that permits prompt evacuation in the event of a fire or other hazard.

emergency spillway See *auxiliary spillway*.

entrance pit An excavated area that serves as the starting point for a horizontal *borehole*.

equator An imaginary line that divides Earth into northern and southern hemispheres.

equipment cabinet An enclosure for equipment to protect against weather and vandalism.

equipotential Having the same *voltage* at every point.

evacuation corridor A portion of a *tunnel* that can be used for emergency egress.

evaporation The conversion of a liquid to a gas.

excavator A construction machine consisting of a *boom*, *bucket*, and rotating *cab*.

exhaust Air that is deliberately removed from a *tunnel* or building.

exit gate A gate on the exit side of a railway *grade crossing* used to discourage motorists from bypassing the warning devices.

expansion joint A gap between constrained structural elements meant to accommodate expansion or contraction.

expansion loop A slack section of *CATV* line that allows for thermal movement of the cable.

extend The act of increasing the length of a *telescoping boom*.

F

facing panels Exterior elements attached to the face of *retaining walls* that protect against erosion, serve as an attachment for *anchors*, and improve the wall's appearance.

fall protection equipment *Personal protective equipment* meant to minimize injuries in the event of a fall.

fatberg A *sewer* blockage created by the accumulation of any combination of fats, oils, grease, wipes, and rags.

feeder The lines in electrical power distribution that connect a substation to distribution transformers.

feedhorn A funnel-shaped *antenna* used to focus high-frequency signals.

feed line The cable that connects a radio transmitter to an *antenna*.

fence A structure that encloses an outdoor area, usually as a safety or security measure.

fender A protective device used between a ship and a dock or another ship.

fiber-optic cable A flexible, transparent cable used to transmit light as a means of digital communication.

field of view The visible area of a person's immediate environment.

figure-8 cable A type of cable used for outdoor aerial applications that includes both the telecommunication line and messenger cable in a single jacket.

fill An area where material is added in an *earthwork* project.

filter The part of a subsurface drainage feature that excludes soil particles from migrating through the *drain*.

filter module An individual, replaceable filter unit.

filter sock A tubular erosion control device installed along the surface to reduce the velocity and sediment load in stormwater *runoff*.

filtration The process of passing *raw water* through media to separate the water from unwanted particles.

final grade The desired final level of the ground in an *earthwork* project.

finished water reservoir An open pond used to store *potable* water.

fire hydrant A connection point to the *water distribution system* used by firefighters.

fish ladder A structure meant to enable fish to swim upstream around a *dam*.

fishplate A bracket used to attach two lengths of *rail* together.

fishway See *fish ladder*.

fission A reaction that splits the nucleus of an atom into two or more lighter atoms.

fixed crane A *crane* installed in a specific location on a *construction site* that cannot move.

flange (beam) An element of a *beam* used to resist bending stress, usually connected via the *web*.

flange (rail) The projecting rim of a wheel used to prevent a car from sliding off the *rail*.

flap gate A *gate* that allows water to flow in only one direction.

flare An open flame used to combust an unwanted gas.

flashover See *arc*.

flight A closely spaced series of navigational *locks*.

flip bucket A hydraulic energy-dissipating structure that diverts a stream of water into the air.

floating bitt A device to which ships are moored within a *lock* that can travel with the water level up or down.

floc A loose clump or group of solid particles.

flocculant A chemical that causes individual suspended particles to accumulate into groups.

floodplain An area of land with high vulnerability of being inundated during a flood.

floodwall A linear construction used to hold back floodwaters along a river or coast.

flue gas Exhausted gas from combustion power plants.

fluted-column tank An elevated water tank whose tower consists of a grooved steel *pedestal*.

flyover A roadway *bridge* that connects two *highways* at an *interchange*.

FM radio Transmission of information by radio waves where the frequency of the signal is varied in proportion to that of the message.

foot The bottom horizontal part of a *rail*.

footing A structural *foundation*, usually intended to transfer vertical forces of a wall into the subsurface.

force main A pressurized *pipeline* that conveys wastewater from a *lift station*.

formation A distinct geologic layer of soil or rock.

foundation The part of a structure that connects it to the ground.

freeboard The vertical distance between the water level and the *crest* of an impounding structure.

freight train A group of cars hauled by one or more locomotives on a railway.

frog A device that enables train wheels to cross to another track.

fuel handling building At a nuclear power plant, the building that houses equipment and areas for handling and storing nuclear fuel.

furnace A device used to produce heat; used together with a *boiler* to create steam from liquid water.

fused cutout A device that serves as both a switch and fuse used on primary electrical lines to protect and isolate *distribution transformers*.

G

gallery A horizontal *tunnel* installed in a *dam* to allow for inspections and drainage.

gantry crane A *crane* that straddles an area and is often wheeled.

gas insulated switchgear Switches, fuses, breakers, and other switching equipment used in electric power stations and *substations* that are encapsulated and surrounded by *sulfur hexafluoride* gas for insulation.

gasket A pliable material used to seal the gap between two parts or objects compressed together.

gate (railway grade crossings) A slender bar that drops across the roadway at a railway *grade crossing* to warn of an approaching train.

gate (water) A moveable barrier used to regulate the flow of water.

gauge The distance between the two *rails* in a railway.

gearbox An enclosure with a set of gears that convert the rotational speed and *torque* of an input shaft to an output shaft.

generation The first stage in electric power delivery that involves creating electric power through various means.

generator A machine that converts mechanical energy into electrical energy.

geogrid Gridded plastic strips or fibers used as reinforcement in soil structures.

geostationary satellite An orbiting object whose *orbital period* is equal to Earth's rotational period so it always appears in a fixed position in the sky.

geotextile Fabric used in construction and *earthwork* to filter, separate, or reinforce layers of soil.

girder A horizontal structural *beam*.

glass beads Transparent spheres used to create retroreflective surfaces.

Global Positioning System (GPS) A *satellite*-based navigation system.

grade The slope of the ground or a roadway.

grade crossing An *intersection* where a roadway crosses a railway at the same level.

grade crossing number A unique identifier given to each railway *grade crossing*.

grader A wheeled machine equipped with a small blade used to perform fine grading during *earthwork*.

gravel An earthen material consisting of small rocks.

gravel pack A layer of rocks installed between a *borehole* and well screen that facilitates flow into the *well*.

gravity dam A *dam* that uses its own weight to resist destabilizing forces.

gridded A configuration of *water main* pipes where water can take multiple paths to the same destination.

gridlock A traffic jam that affects multiple *intersections* within a traffic network, bringing large areas of traffic to a standstill.

grinder A machine that cuts or abrades solids into small pieces.

grit Heavy solids, such as *sand* and soil, found in a wastewater stream.

grit chamber A basin used in the *primary treatment* of wastewater to remove grit from the stream.

groin A structure perpendicular to the shore used to protect a beach from erosion.

ground freezing A method of *dewatering* excavations by freezing a saturated layer of earth to create an impermeable barrier.

ground grid An array of conductive elements used to create an *equipotential* between equipment and the earth.

grounding electrode A conductive element used to make an electrical connection to the earth.

ground-level tank A water storage tank installed on or near the ground.

grout A thin material used to fill small spaces, often including cement to harden over time.

guard rail (railways) A short piece of *rail* parallel to the main rail that helps prevent derailment at *switches* and sharp curves.

guardrail (roadways) A safety *barrier* intended to prevent an errant vehicle from colliding with a roadside *obstacle* or leaving the roadway in a dangerous location such as a cliff.

guide signs Signs that assist drivers in navigation to a destination.

gutter A shallow *channel* used to carry *runoff*.

guy A cable used to stabilize a freestanding *tower* or pole.

guyed tower A vertical structure that depends on *guy* lines for support.

H

hanger A vertical cable that supports a bridge *deck* from above.

harbor An area of still and deep water used for anchoring boats and ships.

hard hat A helmet used on *construction sites* to minimize injuries from bumps and falling objects.

head The top part of a *rail* on which wheels ride.

head-end A facility that receives signals for local distribution in a *CATV* network.

headlight A light or lights at the front of a vehicle.

headwall A wall that supports the end of a *culvert* and directs flow into the pipe.

high-service pump A *pump* used to pressurize a *water distribution system*.

high-visibility clothing Apparel equipped with bright colors and reflective stripes to improve visibility of workers on a *construction site*.

high-voltage direct current (HVDC) A type of electrical power transmission that involves converting the standard *alternating current* of the grid to *direct current* at the beginning of the line and back to AC at the end of the line.

highway See *controlled-access roadway*.

hitch A complication of a *sling* used to secure a load to a *crane* or *hoist*.

hoist A machine used to lift or lower a load.

hook The device at the end of a *crane* cable on which *rigging* and loads hang.

hopper A device or depression that is usually conical in shape and is used to collect or store solids.

horn An audible warning device mounted to train engines and operated to alert people and animals.

hot A *conductor* that is energized above the ground potential.

hub The central part of a rotating device where spokes or blades are attached.

hustler See *terminal tractor*.

hydraulic grade line The surface of water in an open *channel* or the level to which an open vertical pipe would fill when connected to a pressurized *pipeline*.

hydraulic jump A hydraulic phenomenon that forms when a high-velocity stream transitions to a slower velocity, creating a turbulent standing wave.

hydrostatic pressure *Pressure* exerted by a fluid at equilibrium.

I

ice bridge A structure that protects antenna *feed lines* from falling ice.

ice lens A bulging formation of ice that can occur when water in the subsurface freezes.

immersed tube construction A method of building *underwater tunnels* that involves sinking and connecting prefabricated sections.

impact basin A structure used to dissipate the hydraulic energy of water flowing in a pipe.

impact head A device at the end of a roadway *guardrail* that slides along the rail to absorb the impact of a collision and direct the rail away from the vehicle.

impeller The spinning element in a centrifugal *pump*.

induced crack A *crack* that forms along a weakened *control joint* to reduce the occurrence of random cracking in a *concrete* structure.

inductive loop sensor A traffic-detecting sensor that uses a coil of wire embedded in the roadway.

inflow and infiltration (I&I) The unwanted entrance of stormwater and groundwater into a *sanitary sewer* system.

injection system Equipment used to introduce chlorine disinfectant to a *potable* water stream.

inspector A person who verifies that construction is performed in accordance with applicable plans, specifications, and codes.

instrument transformer A device used to isolate sensitive monitoring and control circuitry from the high *voltages* or *currents* of the *electrical grid*.

insulator A device or material that resists the flow of electric *current*.

intake A structure used to collect water from a river, lake, or ocean.

interceptor sewer The largest category of sewer line that collects wastewater from *trunk sewers* and flows to a *wastewater treatment plant*.

interchange A junction of two roadways that relies on grade separation to reduce interruptions.

interior wall A dividing wall within a tunnel that isn't part of the *lining*.

Internet of Things (IoT) Physical objects embedded with sensors and having the ability to exchange data over the internet.

intersection The overlapping area where two or more roadways cross.

inverted siphon A *tunnel* or *pipeline* configuration where a portion of the *conduit* runs deeper below the ground and flows full under *pressure*.

isolation switch See *disconnect switch*.

J

jackhammer A vibratory tool used to break apart rock, *concrete*, *asphalt*, and other hard materials.

Jersey barrier A modular concrete *barrier* used to separate lanes of traffic.

jet fan A fan mounted inside a *tunnel* to induce ambient airflow.

jet pump A *pump* that uses a high-velocity jet to draw a fluid upward.

jetty A structure that projects into the ocean to protect a *harbor* or *canal*.

jib An extension to a *crane boom*.

joint An intentional discontinuity in a structure or structural element.

joint pole A *utility pole* shared by more than one utility provider.

jumper A short length of wire used to connect incoming and outgoing wires.

K

keeper A dangerous hydraulic phenomenon that tends to retain objects in the flow.

kilo-Volt-Ampere (kVA) The unit used for electrical power in *alternating current* (AC) circuits.

kilowatt (kW) The unit used for electrical power in *direct current* (DC) circuits or *alternating current* (AC) circuits where the load is purely resistive.

L

labyrinth weir An overflow structure that is folded into a series of trapezoidal or triangular cycles to provide a greater length for an overall flow width.

lateral earth pressure The *pressure* applied to a retaining wall from the weight of the retained soil.

lateral sewer The *sewer* line that collects wastewater from individual homes and businesses and flows into a *branch sewer*.

lattice tower A *tower* consisting of a framework of structural members assembled in a *truss* formation.

layout The horizontal and vertical configuration of a roadway.

leaf The main element of a water control *gate* that blocks or allows flow.

levee An *embankment* installed along a river *bank* or shore to hold back floodwaters.

level indicator A device used to show the level in a storage tank from the outside.

lift An individual layer of compacted *fill* in *earthwork*.

lift station A structure used to pump *drainage* or wastewater to a higher level.

lightning rod A conductive element mounted in a high location to create a preferential path for lightning strikes, protecting a structure or sensitive equipment.

line A *pavement marking* that delineates lanes, *shoulders*, parking spots, and other roadway features.

line extender See *amplifier*.

lining See *tunnel lining*.

loader A machine equipped with a large *bucket* used to excavate, transport, and load materials.

load following The act of increasing or decreasing electrical power *generation* to match changing demands.

load shedding The act of disconnecting electrical service to groups of customers to reduce total electrical demands on the grid, usually to prevent uncontrolled disruptions or damage to equipment.

local telephone exchange A facility that connects *telephone* lines to establish connections between subscribers.

lock An enclosed structure used for raising or lowering a ship in a *canal* by raising or lowering the water level.

log-periodic antenna A *directional antenna* with multiple elements specifically designed to work with a wide range of radio frequencies.

longitudinal ventilation A *tunnel* ventilation scheme where air flows from one end of the tunnel to the other.

longshore drift The phenomenon of sediment transport along the shore parallel to the coast.

louvers Horizontal openings with angled slats that allow ventilation through an enclosure.

low-Earth orbit An orbit around Earth, generally defined as being within a third of Earth's *radius* above the surface.

low-head dam A small *dam* used to create an elevated and consistent level in the river upstream.

low impact development (LID) The use of processes that mimic natural watersheds to reduce the volume and increase the quality of stormwater *runoff*.

low-noise block (LNB) A device mounted on a *satellite dish* receiver that collects radio waves from the dish and converts them for use in a circuit.

low water crossing A road across a river that is designed to be overtopped and impassable when flows are high.

luff The act of tilting a *crane* boom up or down.

M

machine-bored tunnel A *tunnel* created by a *tunnel boring machine*.

main cable On a *suspension bridge*, the cable that spans between the *towers* and provides the primary support to the *deck*.

main sewer The sewer line that collects wastewater from multiple *branch sewers* and flows into a *trunk sewer*.

mains voltage The electrical service *voltage* delivered to the end user (typically 120V and 240V in the United States).

mangrove forest A species of trees that grow in marshes and along coastlines and have dense, interwoven roots.

manhole A structure that allows personnel access to a stormwater or *sanitary sewer* system.

manually bored tunnel A *tunnel* bored using explosives or excavations and not a *tunnel boring machine*.

marker (fire hydrants) A device used to indicate the location of a *fire hydrant* above the snow.

marker (telecommunications lines) A plastic wrap placed around a *telecommunications* line to identify its type or source.

mast The vertical support structure of a *tower crane* or *satellite dish*.

mechanically stabilized earth Soil constructed with artificial *reinforcement*, often as part of a *retaining wall*.

median A strip of land between opposing lanes of traffic.

membrane A thin sheet of semipermeable material.

messenger wire A structural cable used in aerial applications to support the signal-carrying cable.

microorganism An organism too small to see by eye, including bacteria, protozoa, and some fungi.

microwave antenna An *antenna* used to transmit or receive microwave radio signals.

minor road The lower-capacity road at an *intersection*.

miter gate One of a pair of *gates* used on *canals* or *locks* that hinge at the outside and meet at a point in the center.

mixed liquor The combination of raw wastewater and *activated sludge* at a *wastewater treatment plant*.

mobile crane A *crane* that can move locations on a *construction site*.

moment See *torque*.

monolith A single and continuous block of stone or *concrete*.

monopole antenna An *antenna* consisting of a single conductive element mounted above a conductive surface called a ground plane.

monopole tower A *tower* consisting of a single pole anchored to the ground.

mooring line The rope or chain used to attach a ship to a dock or *wharf*.

morning glory spillway A funnel-shaped *spillway* that projects into a *reservoir* to create a circular *weir*.

mountain tunnel A *tunnel* that passes through a mountain to avoid a route along the surface.

movement One of the many actions that can be taken at an intersection.

multicolumn tank A *water tower* that consists of multiple support legs that hold the tank up.

multiple arch dam A *dam* that uses a series of *arches* supported by *buttresses* along its length.

N

nacelle The streamlined enclosure around a wind *turbine's gearbox*, *generator*, and other internal equipment.

nappe The curtain of water that passes over a *weir*.

natural grade The original surface of the ground before construction.

naval architect One who designs ships or water-borne vehicles.

neckdown A narrowing of a roadway used as a *traffic calming* measure.

neutral A *conductor* that serves as a return path for current in a circuit and is usually at ground potential.

neutral temperature The temperature at which a *rail* is free from thermal stress.

non-ionizing radiation Radiation that does not have enough energy to strip an electron from an atom or molecule.

normal force The contact force between two surfaces.

nozzle cap A protective cover used on the nozzle of a *fire hydrant*.

nuclear power station An electrical power generating plant that relies on a nuclear *reactor* as a source of heat.

nut A device with a threaded hole used in combination with a bolt to create a fastener.

O

object marker A sign that marks a permanent *obstacle* within or beside the roadway.

obstacle Any object or landscape feature that could endanger a vehicle departing the roadway.

obstruction Any object that blocks a driver's view of the roadway.

off-channel reservoir Stored water in an upland area using *dam* that mostly or completely encloses the area.

off-ramp A one-way road leading away from a *controlled-access roadway*.

ogee A curved shape used on a *weir* to improve hydraulic efficiency.

omnidirectional antenna An *antenna* that transmits or receives signals of the same strength in all directions.

on-ramp A one-way road leading onto a *controlled-access roadway*.

operator A device that opens or closes a *gate*.

optical node A device that converts a *fiber-optic cable* signal to radio frequency and sends it over *coaxial cable* lines for distribution to subscribers.

orbital period The time a *satellite* takes to complete one orbit around another object.

outer jacket The protective coating around a *conductor*.

outfall A structure that transitions a concentrated flow of water into a natural watercourse.

outlet A structure that discharges water, *effluent*, or *drainage*.

outlet works A structure or group of structures used to make releases of water from a *reservoir* for downstream use.

outrigger A beam on a *crane* used to increase its stability.

overflow A *pipe* used to release water from a tank in the event it becomes overfull.

overhead telecommunications *Telephone*, *fiber-optic*, or *coaxial cables* installed above the ground and supported by *utility poles*.

overtopping protection Armoring added to an *embankment dam* to protect against erosion from flows that overtop the structure.

P

paddle A small flow control *gate* used to admit or release water from a *lock*.

pad-mounted transformer A *distribution transformer* mounted at ground level in a steel enclosure and used with underground distribution lines.

painted bike lane A *bicycle lane* demarcated only by *pavement markings*.

pantograph A device on an electric locomotive used to collect *current* from an overhead *contact wire*.

parabolic antenna An *antenna* that uses a reflective dish to direct and concentrate radio signals.

parking lane An area adjacent to a *travel lane* on a roadway; meant for vehicle parking.

passing loop See *siding*.

passive warning device A *traffic sign* or *pavement marking* used to warn motorists of danger at a railway *grade crossing*.

pavement The durable outer surface of a roadway usually made from *asphalt* or *concrete*.

pavement marking Paint or thermoplastic applied to the road surface as a warning or guide to motorists.

paving machine A construction machine that lays *asphalt* or *concrete* precisely so it can be compacted into place.

pedestal A small protective enclosure that provides access to *underground telecommunications* lines.

pedestrian cross lights Lights on a *traffic signal* that indicate when it is safe for pedestrians to cross.

pedestrian scramble A *movement* at a *signal-controlled intersection* where all vehicular traffic is stopped and pedestrians are allowed to cross in any direction, including diagonally.

penstock A *conduit* that carries water from a *reservoir* to a hydropower *turbine*.

personal protective equipment Any equipment used by an individual to increase safety or minimize the chance of injury.

phase One of the energized lines in an *alternating current* transmission or distribution *circuit*.

piano key weir An overflow structure that is folded into a series of rectangular cycles to provide a greater length for an overall flow width.

pig A device used to clean the inside of a *pipeline*.

pigtail An accumulation of rags and wipes in a sewer that creates a large, fibrous ball that easily clogs *pumps* or pipes.

pile A vertical structural element drilled or driven into the subsurface; used in *foundations* and *retaining walls*.

pile cap A structural member that distributes loads to one or more *piles*.

pile driver A machine used to hammer or vibrate a *pile* into the earth.

pipeline A *conduit* or tube used to convey a liquid.

pitch The angle of a *blade* relative to the *turbine* axis.

placebo A nonfunctioning device that may provide a perceived benefit.

plain-old telephone service (POTS) Voice-grade *telephone* systems that transmit analog signals over *twisted pairs* of conductors.

platform A structural support that allows mounting of *antennas* to a *monopole tower*.

platoon A group of nearby vehicles traveling in the same direction.

Plimsoll line A reference mark on a ship's hull that shows the maximum depth to which the vessel may be safely loaded.

plunge pool A hydraulic energy-dissipating structure that consists of an armored depression into which discharges plunge.

point The moving part of a railway *switch*.

pole One of the two points where Earth's axis of rotation intersects the surface.

pool A zone of storage within a water *reservoir* with a dedicated purpose.

port (intake) An opening in an *intake* structure through which water can enter.

port (maritime) An area where ships load and unload.

portal The entrance to or exit from a *tunnel*.

post A vertical element that provides support to a *traffic sign*.

potable Safe to drink.

pot bearing A bridge *bearing* that consists of an elastomeric pad mounted inside a steel enclosure.

pothead See *cable termination*.

pothole An unwanted depression in a road surface.

power factor A measure of the synchronization between *voltage* and *current* waveforms in an *AC* circuit.

power rating The maximum power for which a particular piece of equipment was designed.

power transformer A device that transfers power from one *circuit* to another without changing the frequency, usually at higher or lower *voltage*.

precast concrete *Concrete* that is cast in an off-site facility and delivered to a *construction site* ready to install.

preemption device A device that can communicate with emergency vehicles to change traffic *signals*.

preservative A chemical used to prolong the lifespan of wood by preventing natural decomposition by microbes, insects, and fungus.

pressure A physical force exerted continuously over a unit area.

pressurized pipeline A *pipeline* carrying a fluid above the ambient *pressure* outside the pipe.

pre-stressed concrete A *concrete* structure in which the reinforcing steel is tensioned before the concrete cures to increase rigidity.

primary clarifier A circular pond used in *wastewater treatment plants* to settle suspended solids before dissolved nutrients are removed.

primary distribution conductor See *feeder*.

primary electrical distribution Electrical distribution lines on the high-voltage side of *distribution transformers*.

primary treatment The process of removing solids from wastewater as the first step in treatment.

principal spillway A smaller *spillway* at a *dam* that discharges normal inflows to maintain the *reservoir's* level when full.

prismatic elements Reflective elements arranged in a prism configuration to create *retroreflective* surfaces.

profile The vertical layout of a roadway.

project sign A sign placed outside a *construction site* to identify the project, owner, designer, and other details relevant to the public.

pulley A wheel used to change the direction of force in a cable or cord.

pump A device that increases the *pressure* or flow rate of a fluid.

pump column The pipe between the *pump* motor and *impeller* that withdraws water from a *sump* or *well*.

pump house A structure erected around a *pump* to protect equipment and facilitate maintenance.

pumping station A structure consisting of *pumps*, pipes, and other equipment used to lift or deliver a source of water.

pylon See *transmission tower*.

Q

quay See *wharf*.

queue A line of stopped vehicles at a traffic *signal*.

quiet zone A designated stretch of railway where trains are directed not to sound their *horns*.

R

radar detector A vehicle sensor that uses radar for detection as part of an actuated *traffic signal controller*.

radiation pattern The relationship between direction and strength for an *antenna*.

radiator A device used to reject heat to the surrounding air in order to cool a fluid or piece of equipment.

radio antenna tower A vertical structure used to extend the line of sight for mounted *antennas*.

radius The distance between the center and outer edge of a circle or circular arc.

rail Steel bars placed on the ground to form a railroad.

railfan An enthusiast of trains and railways.

railroad tracks A pair of *rails* combined with perpendicular *ties* to form a continuous path for trains.

rail trail A railroad *right of way* that has been converted into a pedestrian trail.

raised pavement marker A safety device attached to a roadway surface used to demarcate *travel lanes*.

rake A device that removes debris from a *trash rack* or *screen*.

rapid transit tunnel A *tunnel* used for rapid transit train systems such as a subway.

raw water Non-potable water taken directly from a source such as a river or lake.

reach stacker A vehicle used in a container *terminal* that can transport and stack *containers*.

reactor A construction used to control a nuclear reaction.

reamer A tool used to enlarge a *borehole*.

recloser A type of *circuit breaker* that automatically reenergizes the circuit after a short delay to protect equipment against transient faults.

recurve A backward curve used on a *seawall* to reflect waves back to sea and minimize overtopping.

reflector A device used to redirect and concentrate radio waves as part of an *antenna*.

regenerative energy Energy that returns to the source or is stored when a motor decelerates or descends.

registration arm The part of a *catenary* electric railway system that holds the overhead *contact wire* in the correct horizontal location.

regulatory sign A *traffic sign* that indicates a traffic rule or law.

reinforcement Material used to strengthen a structure or assembly.

relay A protective device used to trip a *circuit breaker* when a fault is detected.

remote concentrator A device that connects multiple *telephone* lines to a smaller number of switching paths.

remote radio head A device in wireless networking that contains the radiofrequency and signal conversion circuitry.

repeater A device that receives a signal and retransmits it to extend the range of transmissions.

reservoir An area of stored water.

reservoir intake A structure used to withdraw *raw water* from a *reservoir* or lake.

residual The disinfectant remaining in the water at the tap.

resistance A measure of a material's opposition to the flow of electric *current*.

retaining wall A structure that provides lateral support to a soil slope.

retention pond An artificial pond that is normally wet, created to temporarily store rainwater *runoff* to reduce flooding.

retract The act of decreasing the length of a *telescoping boom*.

retreat The strategy of relocating developments at elevated risk of flooding.

retroreflection The returning of light toward the direction of its source.

revetment An armored facing used to protect a shore or *embankment* against erosion.

rigging The act of attaching a load to a *crane* or *hoist*, or the equipment used to do so.

right of way (land) The strip of land immediately below or adjacent to a linear structure or utility such as an electrical *transmission line*.

right of way (traffic) The right of a vehicle to proceed into an *intersection*.

riprap A layer of stones used to protect against erosion.

rise The vertical distance in water levels between the incoming and outgoing *canals* at a *lock*.

rocker bearing A bridge *bearing* that includes a rocking element to provide freedom of movement for thermal expansion and contraction.

rockfill Material used in construction that consists of any combination of *gravel*, rocks, or boulders.

roller bearing A bridge *bearing* that includes a roller element to provide freedom of movement for thermal expansion and contraction.

roller compactor A machine used to compress layers of soil, *gravel*, *concrete*, and other granular materials.

rolling gate A *gate* used in a *canal* or *lock* that rolls along its bottom to open or close.

rolling outage Intentional, temporary disconnection of power to groups of customers in nonoverlapping time periods; used to reduce demands on the *electrical grid*.

rolling stock Any vehicle used on a railway.

roof The upper covering of a *tunnel* or other structure.

rotor shaft The central rotating component of a wind *turbine*.

rough terrain crane A wheeled *crane* that can move to various locations on a *construction site* but cannot travel on a *highway*.

roundabout An *intersection* in which vehicles travel in a single direction around a circular road.

route marker A *traffic sign* that indicates the identifying name or number of a road or *highway*.

rubber joint filler Material used to fill an *expansion joint* in a concrete structure.

rumble strip A tactile warning feature on a *highway* that rumbles when driven over.

running rail The *rail* on which wheels ride.

runoff Water that runs along the ground, usually from precipitation.

rush hour The time or times of day when traffic is heaviest in an urban area.

S

Saccardo nozzle A structure used to deliver fresh air and induce longitudinal airflow within a *tunnel*.

saddle A device used to make a *service connection* on a *water main* pipe.

safety barrier A *barrier* between pedestrian paths and vehicle travel lanes on a *bridge*.

safety space The area below energized power lines on a *utility pole* that provides protection to *telecommunications* technicians from shocks.

sag A vertical curve connecting two inclined sections of roadway at a low point.

sand Soil with particles that are finer than *gravel* and coarser than silt.

sanitary sewer A *pipeline* that carries domestic wastewater.

satellite An object in orbit around a celestial body.

satellite dish An *antenna* used to collect radio signals from *satellites*.

saturated Operating at full capacity.

scaffolding Temporary platforms used to support workers and materials during construction.

scissor lift An *aerial lift* that uses a series of linked and crisscrossed supports to position workers in high or difficult locations.

scraper (clarifier) A device that moves along the bottom of a *clarifier* to push *sludge* into a central *hopper*.

scraper (earthwork) An earth-moving machine that excavates and transports soil using a horizontal blade and pan.

screen A mesh of bars or wires that retains debris while allowing liquid to pass through.

scrubber A device used to reduce air pollution, often using a liquid spray.

scum Floating solids in wastewater.

seawall A structure along the shore used to protect coastal areas from storm surge and high waves.

secondary clarifier A settling tank used after *primary treatment* in a wastewater treatment plant to separate *effluent* from *activated sludge*.

secondary electrical line Electrical *distribution* lines on the low-voltage side of *distribution transformers*.

secondary treatment The removal of nutrients from sewage in a *wastewater treatment plant* after settleable solids have already been removed.

secondary voltage See *mains voltage*.

sector antenna A directional microwave *antenna* often used in cellular communication *base stations*.

sector gate One of a pair of *gates* used on *canals* or *locks* that are shaped like circular sectors and hinge at their centers to meet in the middle of the waterway.

sedimentation The process of removing solids from a wastewater stream using gravity.

seepage The flow of groundwater beneath or along a structure.

self-supporting tower A vertical structure that does not depend on *guy* lines for support.

service connection The pipe that connects an individual customer to a *water distribution system*.

service spillway See *principal spillway*.

sewer A pipe that carries away unwanted waters.

shaft A device that transmits *torque* from a motor to a pump *impeller*.

sharrow A *pavement marking* that indicates what part of the road should be used by cyclists.

sheepsfoot drum A roller used on a *roller compactor* that has numerous lugs or bumps to increase compaction of fine-grained soils.

sheet pile A slender, wide *pile* meant to interlock with its neighbors to form a continuous subsurface wall.

shell The outer section of an *embankment*.

shield A temporary structure used to protect workers and equipment during the excavation of a *tunnel*.

shield wire A grounded *conductor* that runs along the top of *transmission lines* to protect energized conductors from lightning strikes.

shipping The act of transporting and delivering goods.

ship-to-shore crane Large *cranes* used to load and unload cargo from ships.

shoe A contact block that gathers current from an electrified *third rail* or *contact wire*.

shore protection structure Any structure designed to combat erosion along the coast.

shotcrete A method of applying *concrete* to vertical or overhead surfaces using compressed air.

shoulder A lane on the edge of a *highway* generally reserved for emergency vehicles or breakdowns.

shutoff valve A valve used to disconnect a *pipeline* from the *water distribution system* for repairs or maintenance.

side pond See *water-saving basin*.

side slope The area from the *bank* to the bottom of a *channel* or the grade of that area.

sidewalk A paved pedestrian path that usually runs parallel to a roadway.

siding A short section of *railroad track* parallel to the main railway used for passing, loading, and unloading vehicles.

sight distance The unobstructed distance that a driver can see ahead of their vehicle.

signal A device that controls the flow of traffic on a roadway or railway using colored lights.

signal bungalow An enclosure used to house railway *signal* and warning device control equipment.

signal-controlled intersection An intersection where a traffic *signal* controls the flow of vehicles.

signal coordination The configuration of multiple traffic *signals* along a single roadway to work in conjunction to control the flow of traffic.

signal head The part of a *signal* that houses the lights.

sign bridge A *traffic sign* support structure that spans an entire roadway and is supported with vertical elements at both ends.

sign-controlled intersection An intersection where *traffic signs* control the flow of traffic.

silt fence A short erosion control fence usually installed along the perimeter of a *construction site* to reduce the velocity and sediment load in stormwater *runoff*.

single-pedestal tank A *water tower* that uses a single steel column to support the elevated steel tank.

sinker A weight used to hold a *buoy* in place in a waterway.

skid-steer A small construction vehicle often used as a *loader* with a *bucket*.

skimmer A device that collects and removes *scum* in a wastewater stream.

skin effect The tendency of an alternating electric *current* to flow along the surface of a *conductor* instead of the entire cross-sectional area.

slack loop An additional length of aerial *fiber-optic cable* used to facilitate splices or repairs.

slew The act of rotating a *crane* around a vertical axis.

slide gate A water control *gate* that slides within guides to open and close.

sling A length of rope, cable, chain, or webbing used to attach a load to a *crane* or *hoist*.

slip base A joint used for *traffic sign posts* that provides the capability of the sign to break away in the event of a vehicle collision.

slipforming A construction method that places *concrete* into a mold that continuously moves to create linear structures like *curbs* and *barriers*.

slope A surface that is not level where one end is high than the other.

slope paving A durable surface, usually *concrete*, placed on a slope to protect against erosion.

sludge Settled solids in a *wastewater treatment plant*.

slurry A mixture of solids and liquids that behaves like a liquid.

soil-cement A mixture of soil, cement, and water often used as armoring on an *embankment*.

soil nail A structural element installed in an earthen slope to provide reinforcement against slope failure or as part of a *retaining wall*.

spacer A device that holds multiple *conductors* of the same phase within a *bundle* on high-voltage *transmission lines*.

spaghetti junction A colloquial term for a highway *interchange* with multiple levels and ramps.

spark gap Two *electrodes* arranged to allow an electric spark to travel across the gap between them.

spectroscopy A method of identifying chemical constituents by measuring the absorption of different frequencies of light.

speed bump A raised area of a roadway used as a *traffic calming* measure, usually in parking lots and garages.

speed hump A raised area of a roadway, wider than a speed bump, used as a *traffic calming* measure on streets.

speed limit The regulatory maximum speed that vehicles can travel on a particular segment of roadway.

speed lump A raised area of a roadway used as a *traffic calming* measure on streets that has gaps for emergency vehicle tires to pass through.

spigot A shaped area at the end of a pipe that fits into the *bell* of another pipe to connect the two.

spike A large nail used to secure a *rail* to a *tie*.

spillway A structure or group of structures used to make releases of water to maintain the level of the *reservoir*.

splice enclosure A case that protects cable splices from damage due to weather.

splicing truck A vehicle outfitted with equipment for making splices in *fiber-optic cables*.

split-phase A type of electric power service that provides two *alternating current* lines that are 180 degrees out of phase with each other and a common *neutral* line.

spool A cylindrical device on which cable is wound.

spreader A device used by cranes and vehicles to lift shipping *containers*.

stabilized entrance Stones or other hard material used at a *construction site* entrance to reduce the volume of mud carried out on vehicle tires.

stack A structure used to vent concentrated gases high above the ground surface to promote dispersion from human activity.

stacker A machine used to move *coal* and other bulk materials into or out of a *stockpile*.

stack interchange A multilevel, grade-separated highway *interchange* where each turn uses ramps to provide a relatively direct connection to the desired direction.

staging area An area in a *construction site* where materials and equipment are stored.

stake A post or stick that is driven into the ground.

standpipe A tall, slender, ground-level storage tank for water.

startup The time between when a traffic *signal* light turns green and when the intersection becomes *saturated*.

static pole A freestanding structure in an electrical *substation* that protects equipment from lightning.

stationing A measuring system used in engineering and construction for locating a distance along a centerline or horizontal axis.

stay A diagonal cable that connects the bridge *deck* to a *tower* on a *cable-stayed bridge*.

stealth cell site A cellular *base station* disguised or designed to blend into the surrounding environment.

stem A part of a slide *gate* that connects the *leaf* to the *operator*.

step-down The conversion of high-voltage electricity to a lower voltage level using a *transformer*.

stilling basin A structure used to dissipate hydraulic energy at the bottom of a *spillway*.

Stockbridge damper A device consisting of two weights suspended by short cables used to reduce mechanical vibrations from wind in overhead *conductors*.

stockpile A supply of material held in reserve in a loose stack.

stock rails The nonmoving *rails* in a railway *switch*.

stoplog slots Slots into which *beams* can be dropped to adjust the upstream water level or *dewater* a downstream structure.

storage bracket A device used to store surplus lengths of aerial *fiber-optic cables* or change the direction of the cable while maintaining an adequate bend radius.

storage silo A structure used to hold bulk materials.

storage trailer A portable enclosure used for secure storage on a *construction site*.

storm sewer A pipeline that carries away *runoff*.

straddle carrier A cargo-carrying vehicle that carries freight underneath a mobile gantry frame.

strain insulator An electrical *insulator* used in tension to withstand the pull of a suspended wire or cable.

strand A single element of many used to form a cable.

streetlamp A light used to illuminate a roadway or area at night.

stringline A line of string used to mark the precise location of a structure or *earthwork* between *stakes*.

stripe A *pavement marking* that separates travel lanes on a roadway.

subgrade The natural earth below a structure or roadway.

submersible pump A *pump* meant to operate below the level of the fluid.

substation A facility containing switchgear, *transformers*, and other equipment used to connect and control parts of the *electrical grid*.

substructure The portion of a *bridge* that transfers loads into the ground including *piers*, *abutments*, and the *foundation*.

sulfur hexafluoride (SF$_6$) A dense gas used as an *insulator* in electrical switchgear.

sump A depression or impoundment that holds water for pumping.

sun kink Buckling of railway *rails* caused by over-heating and thermal expansion.

superelevation The difference in elevation between the banked outer edge of a roadway's horizontal curve and its inner edge.

superintendent The person in charge of super-vising a construction project.

superstructure The portion of a *bridge* that spans a distance, including *girders* and the *deck*.

supply air Fresh air delivered into a building or *tunnel*.

surface water Any water accessible at the surface of the Earth, including streams, rivers, lakes, and oceans.

surge tank A tank that absorbs fluctuations in *pressure* to protect pipes and equipment from damage.

suspension bridge A *bridge* that uses two *main cables* suspended between *towers* to support the weight of the deck.

suspension tower A support for overhead power lines that does not resist significant horizontal tension forces from the *conductors*.

switch An assembly that allows trains to be diverted from the primary direction onto a sec-ondary railway.

switch machine An electromechanical device that operates a railway *switch* in lieu of a manual operator.

switchstand A device used to manually operate a railway *switch*.

T

T1 A communications technology that allows transmission of digital data over a *telephone* line.

tactile pavement Textured indicators installed on stairs, *sidewalk* ramps, and other hazardous locations to warn pedestrians who are vision impaired.

tag line A length of cable or rope used to stabilize a *crane* load from rotating or shifting.

Tainter gate A radial water control *gate* that hinges at each side and is raised and lowered using a *hoist*.

tanker A ship that carries liquid goods.

tap A device that provides multiple connection points to at *CATV* feeder for individual *service drops*.

telecommunications The transmission of informa-tion over long distances using various technologies.

telephone A device that allows for conversation across long distances.

telescoping boom A multipart crane *boom* that can *extend* or *retract* in length.

temporary road A roadway constructed as part of a *construction site* that will be removed when the project is complete.

tension tower A support for overhead power lines that can resist *conductor* forces, even when unbalanced such as at terminations or changes in direction.

terminal Part of a *port* where a specific type of goods is loaded or unloaded.

terminal tractor A truck used to move trailers and shipping *containers* within a cargo *yard*.

thalweg The line connecting the lowest part of a *channel* along its length.

thermal power station An electricity-*generating* facility that uses heat to create steam and drive a *turbine* generator.

thermoplastic A plastic that becomes pliable at elevated temperatures and is solid at normal temperatures.

third rail An additional *rail* in a railway that sup-plies electric *current* to the locomotives.

through arch bridge An *arch bridge* where the *deck* is supported from below the *arch*.

through street A street that connects at both ends and whose traffic has priority over intersect-ing roadways.

through truss On a *truss bridge*, a *truss* that runs both above and below the road *deck*.

thrust The horizontal force generated when an *arch* supports a vertical load.

tie A perpendicular support for *rails* in a railroad.

tieback See *anchor*.

tied arch bridge An *arch bridge* that includes a tension member between the two ends of the arch to balance *thrust* forces.

tie plate A bracket that transfers and distributes the weight of a *rail* to the *tie*.

toe berm A filled area along the downstream toe of an *embankment* to improve its stability.

toilet-to-tap See *direct potable reuse*.

torque The rotational equivalent of a linear force that equals the product of a force and its perpendicular distance from an axis.

tower A tall structure used to support or elevate a device or assembly.

tower crane A fixed *crane* that consists of a *mast* and rotating *jib*.

track circuit An electric circuit used on *railroad tracks* to sense whether a train is present or not on a specific stretch of railway.

tracks A continuous loop of treads or plates used in lieu of tires to propel construction vehicles.

traction motor An *electric motor* used to propel a vehicle.

traffic calming Measures taken to reduce traffic speed or traffic volume.

traffic control devices Markers, *signs*, and *signals* used to guide and control traffic.

traffic sign A sign that conveys information or rules to drivers.

traffic signal controller A computer that controls the lights of a traffic *signal*.

training wall A wall along the side of a *spillway chute* used to contain the flow.

transformer See *power transformer*.

transmission The intermediate stage in electric power delivery that involves moving power from generating facilities to population centers.

transmission line A system of *conductors* used for the bulk transport of electric power.

transmission tower A structure used to support overhead *conductors* in a *transmission line*.

transmitter building A building that houses transmitters and other equipment near a *radio antenna tower*.

transposition tower A *transmission tower* that shuffles the relative positions of each *phase* in a *transmission line*.

transverse ventilation A *tunnel* ventilation scheme where air flows in ducts and is supplied or exhausted at discrete locations along the tunnel's length.

trash rack A screen used to exclude debris from a *spillway* or *outlet*.

travel The act of moving a crane *trolley* inward or outward along a horizontal *jib*.

travel lane The area of a roadway designated for a single line of vehicles to travel.

trench A linear excavation often used to install *underground utilities*.

trencher A construction machine designed to excavate narrow linear *trenches* for installation of subsurface pipes or utilities.

trolley A mechanism on a *tower crane* that rides along the horizontal jib to position the hook.

truncated dome A surface texture used in *tactile paving* as a detectable warning for pedestrians who are vision impaired.

trunk sewer The *sewer* line that collects wastewater from *main sewers* and flows into an *interceptor*.

trunnion A cylindrical projection that acts as a support and hinge.

truss An assembly of structural members that create a stiff, lightweight frame.

truss bridge A *bridge* that uses a *truss* to support the weight of the deck.

tunnel An excavated pathway below the surface of the Earth for water or transportation.

tunnel boring machine (TBM) A machine that bores circular excavations through the earth to create *tunnels*.

tunnel lining A structural support system used to hold open a tunnel against ground *pressure* and reduce infiltration of groundwater.

turbidity Cloudiness within water, usually from suspended solid particles.

turbine A machine that converts wind or steam power into rotational power along a *shaft*.

turnout See *switch*.

turntable The part of a *crane* that allows the *boom* or *jib* to rotate.

twilight wedge The shadow of the Earth that is visible just before sunrise or just after sunset.

twisted pair A set of two wires in a circuit that are twisted together to reduce electromagnetic interference.

twist lock A device that mates to a shipping *container corner casting* for lifting, moving, and securing.

U

ultraviolet light A device that uses ultraviolet radiation to deactivate *microorganisms*.

underground aqueduct A subsurface *pipeline* or *tunnel* used to carry water over long distances.

underground telecommunications *Telephone*, *fiber-optic*, or *coaxial cables* installed below the ground.

underwater tunnel A *tunnel* that runs below a waterbody such as a lake or river.

uniformity The design concept that safety can be improved by making *traffic control devices* consistent and easy to interpret.

uplift Upward *pressure* along the bottom of a structure.

utility line Any linear installation of pipe, cable, or wire.

utility pole A post used to support overhead electric *distribution* lines, *telecommunication* lines, and related equipment.

V

vacuum breaker A *circuit breaker* in which contacts are housed in a vacuum chamber to minimize the formation of electric *arcs*.

valve A device that controls the flow of fluid within a pipe.

valve key A tool used to open or close a subsurface *shutoff valve*.

vent An opening used to prevent buildup of *pressure* within an enclosed area or allow fresh air to flow.

verge The area between a roadway and *sidewalk*.

vertical turbine pump A *pump* that uses a vertical *shaft* to drive submerged *impellers* that push water up a *column pipe*.

viaduct A long *bridge* that carries a roadway or railway over wide depressions or other obstacles in the landscape.

voltage A measure of electric potential between two points, one of which is often the Earth.

voltage regulator An electrical *transformer* that makes small adjustments on distribution *feeders* to maintain voltage within a prescribed range.

voltage transformer A type of *instrument transformer* that scales large values of voltage to small values that can be measured using instruments and *relays*.

vortex A rotational hydraulic phenomenon that allows air to dip below the normal water surface.

vortex breaker A device that redirects flow in a *pump station* to prevent the formation of a *vortex*.

W

walkway A designated area for pedestrians along a *bridge*.

wall A vertical structural element used to divide or provide lateral support.

wand A part on a paving machine that rides along a *stringline* to control the steering and forms.

warning light A flashing light at the top of a *tower* used to increase its visibility to aircraft.

warning marker A ball-shaped device attached to a *transmission line conductor* to make it more visible to aircraft and other human activities.

warning tape A flexible ribbon used to mark the location of underground utilities.

warning time The length of time between *active warning devices* beginning to operate and the arrival of a train at a *grade crossing*.

warrant A set of instructions issued to a train that authorizes specific movements.

wastewater treatment plant A facility that cleans and disinfects wastewater to make it safe for discharging into the environment.

water distribution system A network of pipes, tanks, and *pumps* used to distribute *potable* water to a service area.

water hammer A *pressure* spike that results from a rapid change in velocity of a fluid in a pipe.

water main A primary pipeline within a *water distribution system* to which service connections are made.

water meter A device that measures the volume of flow in a pipe over time.

water-saving basin A small *reservoir* beside a *lock* to store a portion of discharged water so that it can be used to partially fill the lock when needed.

water tower An elevated water storage tank.

water treatment plant A facility that cleans and disinfects *raw water* to make it safe for human consumption.

wearing course The surface layer of a roadway.

weathervane The act of *slewing* a *crane* to minimize wind forces on the structure.

web The part of a *beam* that resists shear forces and connects the *flanges*.

weir A structure designed to allow water to flow over its top.

well An excavation used to withdraw groundwater.

well development The process of cleaning a *well screen* and establishing a well's hydraulic connection to the *aquifer*.

wellhead The aboveground elements of a *well*.

wet well A belowground enclosure used to temporarily store wastewater as part of a *lift station*.

wharf A shore structure to which ships are moored for loading and unloading.

whistle post A sign along a railway that indicates when a train should sound the *horn* ahead of a *grade crossing*.

winch A pulling or lifting device that consists of a cable or chain wrapped around a drum and turned by a crank.

wind farm A group of wind *turbines*.

wind sensor A device used to measure the direction and speed of the wind.

wingwall A wall that separates an *embankment* from the end of a *culvert* and directs flow into the pipe.

Y

Yagi antenna A multi-element *antenna* designed to be highly directional.

yard A temporary storage area at a *terminal* shipping facility.

yaw Movement about a vertical axis.

Index